SUSTAINABILITY IN MARKETING PRACTICE

Strategies for Industry 4.0

AAP Advances on Sustainable Marketing Practices

SUSTAINABILITY IN MARKETING PRACTICE

Strategies for Industry 4.0

Edited by

Subrata Chattopadhyay, PhD
Udit Chawla, PhD

APPLE ACADEMIC PRESS

First edition published 2025

Apple Academic Press Inc.
1265 Goldenrod Circle, NE,
Palm Bay, FL 32905 USA

760 Laurentian Drive, Unit 19,
Burlington, ON L7N 0A4, CANADA

CRC Press
2385 NW Executive Center Drive,
Suite 320, Boca Raton FL 33431

4 Park Square, Milton Park,
Abingdon, Oxon, OX14 4RN UK

Library and Archives Canada Cataloguing in Publication

Title: Sustainability in marketing practice : strategies for industry 4.0 / edited by Subrata Chattopadhyay, PhD, Udit Chawla, PhD.
Names: Chattopadhyay, Subrata (Economist), editor. | Chawla, Udit, editor.
Description: First edition. | Includes bibliographical references and index.
Identifiers: Canadiana (print) 20240350383 | Canadiana (ebook) 20240350405 | ISBN 9781774915882 (hardcover) | ISBN 9781774915899 (softcover) | ISBN 9781032668123 (ebook)
Subjects: LCSH: Green marketing. | LCSH: Industry 4.0. | LCSH: Consumption (Economics)—Environmental aspects. | LCSH: Social responsibility of business.
Classification: LCC HF5413 .S87 2025 | DDC 658.8/02—dc23

Library of Congress Cataloging-in-Publication Data

CIP data on file with US Library of Congress

ISBN: 978-1-77491-588-2 (hbk)
ISBN: 978-1-77491-589-9 (pbk)
ISBN: 978-1-03266-812-3 (ebk)

About the Book Series: AAP Advances on Sustainable Marketing Practices

BOOK SERIES EDITORS:

Justin Paul, MBA, PhD
Professor, University of Reading, England; Professor, University of Puerto Rico, USA; USA Chief Editor, *International Journal of Consumer Studies*

Sahadev, MBA, PhD, PDF
Professor, Marketing and Responsible Enterprise, College of Business, Technology and Engineering, Sheffield Hallam University, UK

EDITOR:

Dipak Saha, PhD, MBA, PDF (corresponding)
Professor, Marketing and Leadership Discipline, Department of Management, Institute of Engineering and Management, University of Engineering and Management, Kolkata, India;
Series Editor, *Contemporary Management Practices*
Co-editor, *Perspectives in Marketing, Innovation and Strategy*
Co-editor, *Perspective in Finance and Digital Transformation of Business*

Subrata Chattopadhyay, MBA, PhD, PDF
Professor, Institute of Engineering & Management, University of Engineering & Management, Kolkata, India; Mentor of Change, Atal Innovation Mission of Govt. of India

All organizations need practical and actionable strategies to effectively function at the strategic level. This however, requires a deep understanding of strategic marketing tools across different disciplines in different organizations. This book series will delineate a pragmatic approach to marketing, offering managers, researchers, as well as students, a guideline for the formulation and implementation of newly developed marketing strategies as business solutions.

This series will provide comprehensive coverage of the application of various approaches of marketing and strategic practice. It will include contemporary themes of the growing aspects of marketing. It will be a global platform to project various newly developed marketing concepts from diverse fields of marketing to combat the ongoing and future challenges globally. The books in the series will serve as an excellent companion to leading edited books on the subject and will help marketing practitioners in applying tools and 'strategic input' from an innovative perspective to the foreground of practical implementation. Volumes will contain the latest concepts and insight on what researchers and business decision makers need by exploring ideas from contemporary application-based clues in different dimensions of marketing and strategy.

This series, with its comprehensive take, will be ideal as a reference in managerial programs for executives globally. It will serve as an excellent companion to leading edited books on the subject. It will help marketing practitioners in applying tools and 'strategic input' from an innovative perspective to the foreground of practical implementation.

Proposed Book Volume Titles but are not limited to:
Sustainability in Marketing Practice: Strategies for Industry 4.0
Editors: Subrata Chattopadhyay, PhD, and Udit Chawla, PhD

Strategic Marketing Practice
Editors: Philip Kotler, PhD Satyajit Chakrabarti, PhD,
Anupam Bhattacharya, PhD, Dipak Saha, PhD, Rabin Mazumder,
Mrinal Kanti Das, PhD, and Udit Chawl, PhD

ABOUT THE BOOK SERIES EDITORS

Justin Paul, MBA, PhD
Professor, University of Reading, England; Professor, University of Puerto Rico Email: justin.paul@upr.edu

Professor Justin Paul, PhD, is currently an honorary Professor at the University of Reading, England, and as Professor at the University of Puerto Rico, USA. He is also a former faculty member with the University of Washington, USA. He has served as MBA Director and AACSB Coordinator at Nagoya University, Japan, and as Department Chair at IIM-Kerala, India. Dr. Paul introduced the masstige model and measure for branding, CPP model for internationalization, SCOPE framework for small firms, 7-p framework for international marketing, and TCCM framework for literature reviews. He holds/held titles as Distinguished Scholar at IIM-Kerala, Lebanese American University, and SIBM Pune. In addition, he has taught full courses in Denmark, France, Lithuania, and Poland and has served as a keynote speaker at hundreds of conferences that include UVSQ-France, KSMS-Korea, and the Polish Academy, as well as conferences in India. Dr. Paul has been a visiting professor at the University of Chicago, USA; Vienna University, Austria; Fudan University & UIBE, China; and UAB, Barcelona and Madrid, Spain. He has published three best-selling case studies with Ivey & Harvard. He is known as an author/co-author of books such as *Business Environment (4th ed), International Marketing, Services Marketing, Export-Import Management (2nd edition), and Management of Banking & Financial Services* by McGraw-Hill, Oxford University Press, and Pearson respectively. His total downloads exceed over 800,000 times. An author of over 180 research papers in SSCI journals, Dr. Paul has over 75 papers are in A or A star journals (ABDC) & 65+ in UK-CABS 3- or 4-level journals. He has visited over 80 countries as a visiting professor/public speaker. He holds two PhD degrees from the University of Brighton, England, and IIT Bombay. He is currently the Chief Editor of *International Journal of Consumer Studies,* the world's top consumer research journal ranked by Web of Science with an impact factor 10.0 published from UK and USA. Ranked among the world's top 92 highly cited Professors in Business & Economics subject areas, his citations are increasing at record rate 700 per month, with an H-index 69.

Sunil Sahadev, MBA, PhD, PDF
Professor of Marketing and Responsible Enterprise, Sheffield Business School, College of Business, Technology and Engineering, Sheffield Hallam University, UK
Email: s.sahadev@shu.ac.uk

Professor Sunil Sahadev, PhD, is currently Professor of Marketing and Responsible Enterprise at Sheffield Hallam University, United Kingdom. He started his academic career at the Indian Institute of Management, Kozhikode. He subsequently worked at the University of Sheffield, UK, University of Salford and University of Brighton. He is also a visiting faculty at the Vilnius University, Lithuania. He holds a post-doctoral fellowship from the Asian Institute of Technology, Bangkok. He has published widely in highly repute peer reviewed journals like the *Journal of Services Research, Journal of Travel Research, British Journal of Management, Journal of Business Research, Technology forecasting and Social Change, International Journal of Human Resources Management*, etc. He has also authored a widely used textbook on sales and distribution channel management. He has supervised to completion 17 PhD students. His areas of research interest include managing boundary spanning elements, digital marketing, and sustainability. He completed his PhD from the Indian Institute of Technology, Chennai, India.

Dipak Saha (Corresponding) MBA, PhD, PDF
Professor, Marketing and Behavioral Leadership, Department of Management, Institute of Engineering and Management, University of Engineering and Management, Kolkata, India
Email: dipak.saha@iem.edu.in

Professor Dipak Saha, PhD, is currently associated as an experienced professor of marketing and behavioral leadership with a demonstrated history of leadership with the Department of Management, Institute of Engineering and Management, University of Engineering and Management, Kolkata, India. Dr. Saha has over 20 years of corporate and academic experience. He has published several articles in reputed and renowned scholarly journals. He is the co-editor of the book *Perspectives in Marketing, Innovation, and Strategy* with the "Father of Marketing" Prof. Philip Kotler and *Perspective in Finance and Digital Transformation of Business*. He is a talent strategist and consultant and trainer in the field of strategic decision-making. He is the recipient of the Academic Excellence Award 2020 and the Research Excellence Award 2023 from the Academic of Management Professionals.

His areas of interest are marketing and behavioral leadership. Prof. Saha is also the Series Editor of a global book series *Contemporary Management Practices*, published by Taylor & Francis (Routledge).

Subrata Chattopadhyay, MBA, PhD, PDF
Professor, Institute of Engineering & Management, University of Engineering and Management, Kolkata, India; Mentor of Change, Atal Innovation Mission of Govt. of India
Email: subrata.chattopadhyay@iem.edu.in

Professor Subrata Chattopadhyay, PhD, is currently working as Professor at the Institute of Engineering & Management, University of Engineering and Management, Kolkata, India. Adept at innovative teaching practices, he is highly networked within the industry as he continues to practice alongside teaching. He has contributed in more than 30 international journals and has published over 23 papers in seminars and conferences. He has also authored three books: *Computations in Estate Management, Transport Management,* and *Values and Ethics for Engineers and Managers*. He is on the editorial boards of five international journals and is a NEN-DST Govt. of India recognized faculty on Entrepreneurship. He is a Mentor of Change in the Atal Innovation Mission of Govt. of India and has been conferred the prestigious Certified Management Teacher by MTC-Global for his innovative and dedicated teaching and mentoring in 2015. He has successfully chaired and has been invited as keynote speaker at several national and international conferences in South Asia. Dr. Chattopadhyay has been awarded and recognized as the Outstanding Higher Education Leader of the Year by SOE Global Education Awards in 2014 and has been recognized by NFED for his contribution to mentoring startups, teaching, and research for two consecutive years, 2013 and 2014. He was conferred the prestigious MTC Global Top Thinkers Award in September 2015 and Mentor of the Year by HRD India in June 2019. He earned his *MSc, MBA, and PhD degrees at* IIT-ISM Dhanbad, India.

Series Editorial Board Members

- **Prof. Veland Ramadani,** Professor of Entrepreneurship and Family Business at Faculty of Business and Economics, South-East European University, North Macedonia.
- **Dr. Sudhir Rana,** Associate Professor, College of Healthcare Management & Economics, Gulf Medical University, UAE.
- **Dr. Ghulam Ali Arain,** Associate Professor of Organizational Behavior, Suleman Dawood School of Business, Lahore University of Management Sciences, Pakistan.
- **Prof. (Dr.) Mahboob Ali,** Professor, Dhaka School of Economics, Dhaka University, Bangladesh.

About the Editors

Subrata Chattopadhyay, PhD
Professor, Business Administration Department,
University of Engineering & Management, Kolkata, India

Subrata Chattopadhyay, PhD, is presently working as a Professor in the Business Administration Department at the University of Engineering and Management, Kolkata, India. He is an MSc, MBA, PhD from IIT-ISM Dhanbad. He is well-versed in cutting-edge teaching techniques and has extensive industry connections. In addition to instructing, he also practices, and his pupils adore him for it. He has published more than 23 articles at seminars and conferences and contributed to more than 30 international journals. In addition, he is the author of three books: *Values and Ethics for Engineers and Managers, Transport Management,* and *Computations in Estate Management*. As a life member of ISTD, CMA, PRSI, NIPM, NHRD, NFED, and ISABS, he also provides mentorship and consulting services to a number of other organizations.

Udit Chawla, PhD
Associate Professor, Business Administration Department,
University of Engineering & Management, Kolkata, India

Udit Chawla, PhD, is presently working as an Associate Professor in the Business Administration Department at the University of Engineering and Management, Kolkata, India. He is a BTech, MBA, and PhD in Management. He is an accomplished Associate Professor with a track record of employment in the higher education sector. He has expertise in R, SPSS, SAS, PYTHON, etc. He is a strong education professional focused on business analytics, data science, machine learning, and artificial intelligence. He has published more than 30 articles at international conferences and contributed more than 15 to international journals.

Contents

Contributors

Priyank Kumar Agarwal
Techno India University, Kolkata, West Bengal, India

Sudin Bag
Department of Business Administration, Vidyasagar University, Midnapore, West Bengal, India

Debarati Banerjee
PhD Scholar, Faculty of Management Studies, ICFAI University, Jharkhand, India

Mohua Banerjee
IMI, Kolkata, West Bengal, India

Shivaji Banerjee
St. Xavier's College, Kolkata, West Bengal, India

Subhajit Bhattacharya
Department of Management, XIM University, Odisha, India

Suchandra Bose
Assistant Professor, Faculty of Business Administration Department, University of Engineering and Management, Kolkata, West Bengal, India

Debolina Chakraborty
Associate Relationship Manager, Standard Chartered Bank, Kolkata, West Bengal, India

Subrata Chattopadhyay
Department of Management, University of Engineering and Management, Kolkata, West Bengal, India

Udit Chawla
Associate Professor, Faculty of Business Administration Department, University of Engineering and Management, Kolkata, West Bengal, India

Kamalika Dasgupta
Assistant Professor, Faculty of Business Administration Department, University of Engineering and Management, Kolkata, West Bengal, India

Joysri Datta
Assistant Professor, Faculty of Business Administration Department, University of Engineering and Management, Kolkata, West Bengal, India

Rajib Dutta
Amity Business School, Amity University, Kolkata, West Bengal, India

N. Elangovan
School of Business and Management, Christ University, Bangalore, Karnataka, India

Ajay Kumar Ganguly
Department of Management, University of Engineering and Management, Kolkata, West Bengal, India

Hena Iqbal
Adjunct Faculty, Murdoch University, Dubai

M. A. Jabbar
Professor and HOD, Department of CSE (AI & ML), Hyderabad, Telangana, India

Ankita Jain
Department of Management, IIS (Deemed to be University), Jaipur, Rajasthan, India

Manohar Kapse
Symbiosis Center for Management and Human Resource Development, Symbiosis International University, Pune, Maharashtra, India

Vaishali Mahajan
Symbiosis Center for Management and Human Resource Development, Symbiosis International University, Pune, Maharashtra, India

Sudipta Majumdar
Associate Professor, Department of Management, School of Business & Economics, Adamas University, India

Kousik Mandal
Department of Business Administration, Vidyasagar University, Midnapore, West Bengal, India

Ayus Mehta
BBA, The Calcutta Anglo Gujarati College, Kolkata, West Bengal, India

Pratika Mishra
Professor, Welingkar Institute of Management Development and Research, Mumbai, India

Varsha Mishra
MSc Digital Marketing, Royal Holloway, University of London, UK

Dibyendu Nandy
PhD Scholar, Faculty of Management Studies, ICFAI University, Jharkhand, India

Harshrim Pardal
Senior Cloud Analyst-HCM, Oracle Corporation, Bangalore, Karnataka, India

Radhika Pillarisetty
Associate Professor, Presidency Business School, Presidency College, Hebbal, Bengaluru, India

Jeanne Poulose
School of Business and Management, CHRIST (Deemed to be University), Delhi NCR, India

M. Rajkumar
Associate Professor, Faculty of Management Studies, ICFAI University, Ranchi, Jharkhand, India

Swecha Ram
Future Institute of Technology, Kolkata, West Bengal, India

Tripshita Saha
St. Xavier's College, Kolkata, West Bengal, India

Meha Saxena
Department of Management, IIS (Deemed to be University), Jaipur, Rajasthan, India

Deboshree Barman Seal
Department of Business Administration, Vidyasagar University, Midnapore, West Bengal, India

Kirti P. Sharma
Techno India, Kolkata, West Bengal, India

M. K. Sharma
Faculty of Management, IIS (Deemed to be University), Jaipur, Rajasthan, India

Shivam Prem Sharma
Techno India University, Kolkata, West Bengal, India

Vinod Sharma
Symbiosis Center for Management and Human Resource Development, Symbiosis International University, Pune, Maharashtra, India

Jyoti Shaw
MBA, Jagdish Sheth School of Management(JAGSoM), Bangalore, India

Sahil Shaw
BBA, Techno India, Kolkata, West Bengal, India

Shubham Kumar Shaw
BBA, Techno India, Kolkata, West Bengal, India

Aditya Vikram Singh
MBA, Jagdish Sheth School of Management(JAGSoM), Bangalore, India

Amaan Singh
BBA, Techno India, Kolkata, West Bengal, India

Ankita Singh
MBA, Jagdish Sheth School of Management(JAGSoM), Bangalore, India

Ayush Kumar Singh
Future Institute of Technology, Kolkata, West Bengal, India

Harsh Vikram Singh
BCA, Techno India University, Kolkata, West Bengal, India

Sarabjot Singh
Vellore Institute of Technology, Chennai, Tamil Nadu, India

Dey Subhanan
Research Scholar, Faculty of Management Studies, ICFAI University, Ranchi, Jharkhand, India

Abbreviations

AHP	analytical hierarchy process
AVE	average variance extracted
BFSI	banking, financial services, and insurance
CB-SEM	covariance-based SEM
CDC	Centers for Disease Control and Prevention
CNN	convolution neural network
CONF	confidence
CR	composite reliability
CV	computer vision
DCNN	deep convolutional neural networks
DNN	deep neural network
EUA	emergency use authorization
EV	electric vehicle
FDA	Food and Drug Administration
FLI	food label information
GAN	generative adversarial network
HMM	hidden Markov model
HTMT	heterotrait-monotrait
IEA	International Energy Agency
IIP	index of industrial production
IT	information technology
KMO	Kaiser-Meyer-Olkin
KNN	K-nearest neighbor
ML	machine learning
NAA	not at all
NAB	National Automotive Board
NAD	neither agree nor disagree
NLP	natural language processing
NNs	neural networks
OLS	ordinary least square
PGM	probabilistic graphical modeling
PIR	passive infra-red
PLB	private label brands
PLS	partial least squares
PLS-SEM	partial least square equation modeling
PMU	phasor measurement unit
PoS	parts of speech

PRB	population reference bureau
RPi	raspberry Pi
SARS	severe acute respiratory syndrome
SD	strongly disagree
SEM	structured equation modeling
SLR	systematic literature review
SOOS	shelf out of stock
SSD	single shot detection
SVM	support vector machine
TFQ	total food quality
THR	thresholding
TPB	theory of planned behavior
UPI	unified payments interface
VAB	value-attitude and behavior
VO	very often
WTP	willingness to pay

Foreword

Few marketing issues have been as spirited as "sustainable" over the past few years. The current proliferation of green media, goods, services, and marketing has resulted in a sea of misunderstanding and mistrust, which runs the risk of weakening the entire green movement and bringing us back to a time of apathetic consumer behavior.

Consumers' points of view about the effectiveness of green products have changed dramatically, going from being very hesitant to being commercialized across several categories. At the same time, consumers are rebalancing their lifestyles in response to decreased levels of income and spending as a result of COVID-19, climate change, goods shortages, and inflation, all of which has caused a significant amount of anxiety.

Marketers need to recognize the role of limited resources, high environmental costs, and the growing warming of our planet. Marketers need to modernize their procedures for pricing, distribution, branding, and product creation. The purpose of the marketing team is to achieve good profits within a strong policy and practice of sustainability. Companies must adopt the mantra of "reduce, reuse, and recycle" in developing their products, services, and marketing plans. We need to advance green marketing as a core task. Support for a healthy planet should be part of your regular brand messaging.

This book well describes how organizations need to manage their resources and goals in turbulent times.

—Professor Philip Kotler
S.C. Johnson & Son Distinguished
Professor of International Marketing (Emeritus),
Kellogg School of Management,
Northwestern University, USA

Preface

Sustainable actions are now crucial for any and every brand as our world continues to struggle for survival. Each business enterprise has its tribe, and while that tribe selects a sustainable product over its rivals, their impact is significant. To enhance the quality of life for both the prevailing and future generations, sustainable marketing entails creating and advertising goods and services that cater to the demands of both consumers and business users.

Sustainable marketing is more than just making efforts to appear more socially and ecologically concerned and telling customers about it. It has greater depth and significance. Consumers prefer businesses that support sustainability, which is no longer a secret. This book is a manual on sustainable marketing that will show you how to integrate sustainability into your marketing strategy and keep it there. This book will also educate you on how to develop and broaden a sustainable marketing strategy that appeals to the values of the general public.

Chapter 1: Influencing Consumer Decisions through Nudge Marketing: Evidence from the Indian Fashion Industry

Nudge marketing is a type of marketing that uses psychological factors to influence consumers' purchase decisions in a positive way. Nudges are subliminal messages that persuade customers to purchase goods more quickly and in a manner that suits the producers. While Indian marketers are unfamiliar with the word "Nudge," they have been effectively employing this strategy for a considerable amount of time. The following are some of the most prevalent nudge marketing strategies that have been seen in the Indian fashion industry: Product labeling, suggestions or ratings, price fixing, deceptive assessments, and leave intention overlays. In order to influence customer purchasing behavior in this market, fashion companies use a variety of marketing tactics, including nudge marketing. This research intends to explore these tactics in depth.

Chapter 2: Commodity-to-Brand Transformation: A Review and Research Agenda

When consumers believe a product has a high value and is attractive and exciting to them, it becomes branded. Branding commodities has advantages for both consumers and marketers. A good or service is a raw material that is later sold on the market. Marketers work to establish this brand in a situation where there is a lot of competition, in order to stand out from the rest. Brands work to reduce the risk and offer value over what the consumer is willing to pay. A higher economic value means that the customer is more likely to be satisfied with the product. The study found that when a good or service is made available as a result of transformation by either the producer or the consumer, a brand is born.

Chapter 3: Influence of Personality Traits on the Purchase Behavior of Consumers in Kolkata

A person's personality is the unique combination of characteristics that make them unique and determine how they react in and around their environment. This research tries to give a realistic impression of how consumers usually purchase items. There was a lot of research done on the personality factors that affected how consumers purchased goods and services during the COVID-19 epidemic. This has provided a concise overview of the major personality traits of an individual. This chapter discusses how customers make decisions and how businesses can use this information to their advantage.

Chapter 4: A Conceptual Framework to Decode the Engagement and Satisfaction of Banking Consumers in Digital Space: A Review-Based Study

When performance outcomes and expectations are compared, satisfaction is a feeling that is associated with the customer's experience. This study looks at consumer satisfaction with internet banking. The company's external environment is important to the company's success. It makes it easy for customers to find what they're looking for, and also Privacy and Security conscious. There is evidence that customer happiness, behavioral intention, attitude, and loyalty outcomes can be affected by how expectations are fulfilled.

Chapter 5: An Empirical Study to Understand the Parameters Affecting Customers' Satisfaction toward Fashion Brands Following Sustainable Practices

Fashion is only one aspect of being able to fulfill fundamental requirements. It serves as a means of association with a certain group or set of tastes. The fashion industry is headed toward a time when everything is interconnected and there are several fashion subcultures. Textiles can be both aesthetically pleasing and practical, since they may have a variety of impacts on someone.

Chapter 6: Using Sentiment Analysis to Identify Consumers' Emotions in the Hotel Industry

A significant development engine for the Indian service sector is the tourism and hospitality sector. India is a great resource for the future because of its diverse environment and culture. With a 6.9% contribution to the nation's GDP in 2019, India's tourism sector is in the top 10 among 185 nations. By 2027, this business is expected to be worth 125 billion US dollars. The amount of open positions significantly affects the US economy. This study investigates consumers' emotions influencing hotel reviews and how they evolve over time.

Chapter 7: An Introspection on Fame India Scheme in the Light of Non-Incentivized Electric Private Four-Wheelers: Potentials and Possibilities

The FAME Scheme is a government program to help people switch to battery-powered cars as a viable mass transit option. The project has a two-year schedule that starts in 2015 and ends in 2019. The incentives for firms that are involved in the transportation industry are currently only available to those who are using two-wheelers, three-wheelers, or public transportation vehicles. The government has not yet announced plans to fund private four-wheelers. There are some ideas to increase the effectiveness of the FAME Scheme, as certain electric cars are not being manufactured or purchased. This study project examines the need for incentives in the private electric vehicle (EV) industry and provides recommendations for lowering ownership costs for battery-powered electric cars.

Chapter 8: Application and Sustainability of AI and Machine Learning in the Various Stages of the Value Chain in Online Fashion Retail: A Conceptual Review

With only an eye toward establishing market dominance, companies inside the textile and garment trade have indeed been forced to develop solid and reactive supplier relationships which will provide better business performance and to provide customers the greatest value.

The possible uses of machine learning in the fashion market range widely, from design support networks to fashion suggestion systems.

The analysis of artificial intelligence's short encounter in the fashion market demonstrates clearly that these phenomena and the connected industry, a substantial contributor to the global market, are significantly impacted by AI.

Chapter 9: An Empirical Study in Analyzing Customer Preferences and Adoption of Electric Vehicles

Electric cars are just recently beginning to be produced, despite the fact that electric vehicles have not yet had the same level of commercial success as vehicles powered by internal combustion engines. Electric cars are projected to become more popular in the next years due to upcoming battery advances. Electric cars have the potential to have significant environmental benefits if they are produced from sources like nuclear power plants or highly efficient, contemporary producing stations. Benefits to the economy include no exhaust pollution, less reliance on fossil resources, and lower total carbon emissions. In order to broaden the scope of the current life cycle sustainability assessment framework, this study proposes an innovative and useful methodology that takes into account macroscale social, economic, and environmental factors.

Chapter 10: A Study on the Growth Prospects of Q Commerce and Its Role Play in eCommerce Business in Present-Day India

Some quick descriptions of mobile technology predict that extremely rapid penetration and growth processes will take place. This report predicts that mobile technology will increase in popularity in markets with a low penetration rate of mobile technology. Commerce has become a means of product promotion, as well as providing customers with a comfortable experience. People who switch from traditional retail to online shopping often increased the efficiency and cost-effective package delivery system. If this trend continues, it could indirectly relate to Q Commerce. Additionally, it may improve community service by assisting the environment, particularly in the full-service restaurant service industries, and by supporting SMEs with technological developments.

Chapter 11: Key Factors Influencing Customer Preference Regarding Ecotourism from a Data Analytics Perspective

Ecotourism is a concept that encompasses the entire travel experience, paying attention to economic, social, and environmental concerns as well as the needs of host communities and tourists' experiences. Environmental protection, social equity, quality of life, cultural diversity, and a vibrant, healthy economy that creates jobs and prosperity for all should all be taken into consideration when planning tourism. The idea that tourism should be eco-friendly and beneficial to local people is now widely accepted. If properly developed, planned for, and managed, tourism in all its forms has the potential to be sustainably profitable. Organizations that support tourism development are promoting ecotourism practices to lessen the adverse effects of tourism's expanding influence, such as its effects on the environment.

Chapter 12: What Factors Do Zoomers Consider While Buying Private Label Products? A Conceptual Model

Private label products have lately grown in popularity in retail establishments. Brand positioning refers to a company's attempts to create a distinctive and recognizable identity which will attract its target market. The method of creating a product and making sure it is upheld throughout time is known as brand management. The concept of branding involves developing a distinctive and recognized identity that will attract customers. A market strategy's ability to be successful depends just on the logical structure it employs to carry it out. Inside a crowded market, store brand design might assist you in standing out. This shift in consumer desire helped private label products gain popularity across the nation. Buyers of premium brand goods are looking for reasonably priced, high-quality goods. Private label goods are beneficial to customers because they're more inexpensive and of superior quality.

Chapter 13: Exploring Buying Intention and Brand Loyalty of Millennials and Gen Z towards Sustainable Beauty and Personal Care Products Using a Green Mix Approach

India is a very large and growing country with a strong democracy. In the past few years, India has made great progress in many areas, including its economy and its civil rights. Organizations are moving towards sustainable growth, which means that they are implementing green practices across the product's entire life cycle. In addition, they are providing "Green Products" which are equal or better than traditional products. We're looking for products that are environmentally friendly, without causing any harm. Consumers today are more aware of the ways in which firms take into account their social and ethical responsibilities. This study looks at how important it is for customers to understand the concept of sustainable cosmetics and personal care products, and how that affects their buying decisions.

Chapter 14: Ecotourism: A Study of the Relationship between Demographic Factors and Satisfaction among Customers towards Ecotourism Using the Chi-Square Application

Participating in eco-tourism helps preserve the environment and improve the well-being of local people. Critics of the global growth model argue that it may have negative effects on rural populations and may preserve pristine landscapes at the expense of developing countries. Ecotourism is becoming more popular because it offers visitors places to go and things to do that have a minimal impact on the environment. This study looks at consumer attitudes toward ecotourism in light of demographic variables in order to better understand how ecotourism might be viewed by different groups of people.

Chapter 15: Why Do Consumers Behave… The Way They Do! (A Bird's-Eye View on the Heuristics and Biases in Consumerism)

In a competitive market, understanding your customers' buying habits can help you create more successful marketing campaigns and directly address the demands of your consumers. In a cutthroat market, increasing client loyalty is key. By understanding your customers, you can increase your sales. An improved understanding of the complexities of customer behavior may be used to better represent customers in their dealings with businesses. The chapter provides fresh insights on the life course concept and demonstrates how it can be used to investigate consumers in a novel way.

Chapter 16: Impact of Consumers' Knowledge and Use of Food Labels on Willingness to Pay Using SmartPLS

Food Labeling is an effective tool to protect consumer health in terms of food safety and nutrition. Food labels convey information about the product's identity and contents, and on how to handle, prepare, and consume it safely. Food labeling is a powerful way to ensure that people receive the right nutrients from their food. Food labels provide information about the nutritional components of the food and how they are related to one another. They also provide safe handling, preparation, and consumption procedures. Despite the majority of consumers being interested in FLI, its use is hampered by a lack of understanding of food labels. The requirements for nutritional information are not clear to the public.

Chapter 17: Occupancy Monitoring to Prevent Spread of COVID-19 in Public Places Using AI

This study intends to automate the counting of individuals for occupancy monitoring and send a mail warning if the occupancy exceeds the specified threshold in the case of limited occupancy. This project offers a method to capture and replay live video or audio from a camera. This project is achievable with a laptop and a camera to take images and video. To monitor different locations, the system can employ cameras. The population count will be automated as a result of this project, and any changes will trigger alarms. This study outlines a system that will monitor individuals entering and exiting malls and offer information on the current population of the mall.

Introduction

The book discusses the role of marketing in facilitating sustainable behavior. Contemporary marketing education may not prepare students and professionals to use marketing's influence to support sustainability. The shift toward more sustainable business practices is unlikely to be linear; different societies will likely show different sensitivities toward sustainability and, most likely, different behaviors.

- Marketing and sustainability marketing response to the emerging trend of sustainable discourse;
- Origin of interest in sustainability;
- Practice of overlooking the idea of sustainability toward pursuit of short-term business goals.

The book will enhance knowledge about cutting-edge research in sustainability and will be helpful for researchers, academicians, and corporates immensely.

It talks about innovativeness and high-end new research in sustainability. Sustainable Marketing will be an important and much sought-after area post-COVID scenario. Value generation and delivery will be at the top of the mind of consumers.

The core value of sustainability is embedded in the long-term maintenance of the quality of life and contains environmental, economic, and social components. Business organizations have a substantial responsibility for preserving the quality of life for future generations, as a result of the role they play in transforming natural and societal resources into goods and services. The marketing function is a major force in strategic decision-making in contemporary organizations. Contemporary research on sustainability is fragmented between multiple disciplines and areas, with ongoing debates about methodological as well as practical issues. With its arsenal of tools for influencing (managing) consumption patterns, marketing must take responsibility for the sustainable behavior of both organizations and consumers.

The primary aim of this publication is to establish a framework for and give an overview of major ideas that connect marketing, consumption, and sustainability. Marketing and sustainable marketing have responded to the emerging trend of sustainable discourse quite early. This manuscript addresses dominant areas of research on sustainability from the marketing perspective, the origin of interest in sustainability, as well as the practice of misplacing the idea of sustainability in pursuit of short-term business goals.

PART I
Sustainable Consumer Culture

CHAPTER 1

Influencing Consumer Decisions Through Nudge Marketing: Evidence from the Indian Fashion Industry

KAMALIKA DASGUPTA,[1] JOYSRI DATTA,[1] and UDIT CHAWLA[2]

[1]*Faculty of Business Administration Department, University of Engineering and Management, Kolkata, West Bengal, India*

[2]*Business Administration Department, University of Engineering & Management, Kolkata, India*

ABSTRACT

The way consumer buying decisions are framed in consumers' minds while buying a product is broadly known as consumer behavior. Nudge Marketing is a way in which marketers coax a desired behavior in their consumers by engaging them with their promotional tools and techniques. In short, nudges are subtle indicators that aid customers to make purchases that are faster at the same time preferred by the producers.

In recent times, we have seen some massive changes in the fashion industry. Standing in 2022, where trending with the latest fashion trends is quite imperative, fashion companies are constantly striving to keep up with the consumer needs and wants. The Indian fashion industry is ever-changing, and so are the choices and preferences of the customers across all age groups and genders. With the growth in organized retail and online marketing, this segment is flourishing and will continue to do so in the future.

Our study aims to delve into the marketing techniques and strategies through nudge marketing utilized by fashion firms to influence the consumer purchasing process in this segment. This research paper paves the way for

Sustainability in Marketing Practice: Strategies for Industry 4.0.
Subrata Chattopadhyay, Udit Chawla, (Eds.)

analyzing the prominent factors that have the potential to influence the buying behavior of the customer in the retail fashion segment in India through a self-administered questionnaire used for data collection.

1.1 INTRODUCTION

Most of the economic theories are founded upon the core assumption that human beings are rational in a collective sense. Their decisions are based upon the maximization of their gains and minimization of their losses. This assumption wasn't questioned until Tversky & Kahneman (1989) came up with their pioneering work that seriously challenged the 'rational theory of choice.' Later, a much deeper insight on this was made by Richard Thaler (2017) which popularly came to be known as "Nudge Theory." It proves that human beings aren't as rational as they seem. Thaler and other behavioral economists have shown that in reality people's so-called "rational" choice can be manipulated or channeled in a preferred direction through subtle but indirect interventions. This is called "Choice Architecture"[1] and it can be successfully done because human beings, while making their decisions, are unconsciously shepherded by biases and psychological fallacies.

Nudge as a behavioral tool to architect people's choices is becoming increasingly popular across the globe. While many countries including India are applying this mechanism to frame public welfare policies, it has become extensively popular in the marketing field too. Nudge Marketing in its true sense of the term is all about perceiving the psychological attributes that influence consumers' purchasing decisions and using them to direct consumers' decision-making in a favorable way. It might seem a very scheming strategy that makes decision making stressful, but in reality, it is actually the opposite. Nudges are not always manipulative in essence. The modern Nudge Marketing techniques turn decision making into a pleasing customer experience by creating a win-win situation for both consumers and producers. They deal with consumer psyches such as their preferences, values, buying habits, etc., which the brands use to create specific and customized offers that further improve their sales figures.

Nudge Marketing as a concept is very new in the Indian context. Rather it will be appropriate to put it this way that marketers have been using this approach all along in various ways, but the term "Nudge" was not very familiar until recently. As per the recent data[2], the Indian fashion industry has 5% contribution in India's GDP and 14% contribution to the overall index

[1] Refers to the designing of ways in which choices can be presented to people, especially consumers.
[2] https://mospi.gov.in/web/iip/home (accessed on 12 December 2023).

of industrial production (IIP[3]. The textile industry alone is one of the largest contributors to India's export earnings, accounting for approximately 14%[4] of total exports. Besides being an important industry, it also happens to be a dynamic one with a constantly evolving consumer outlook. The fact that India's population composition is exceptionally versatile makes it even more challenging for the brands to market their products and sustain a satisfactory amount of revenue. With the changing dynamics of the industry itself, the marketing strategies also evolve. Traditional marketing strategies do not apply to modern consumer behavior. The essence of nudge marketing is "Libertarian Paternalism" that allows the companies to direct and affect consumer behavior while completely respecting their freedom of choice. This chapter delves into finding out the most important factors that have the potential to nudge consumers' buying decisions of fashion products and also into identifying how the Indian fashion industry, in particular, is using these nudge marketing strategies to influence consumer choices.

1.2 LITERATURE REVIEW

Application of nudge marketing strategies have been noticeably effective, especially, in the field of promoting sustainable fashion. Luxury companies often suffer from the problem of inventory accumulation. H&M has recently reported an inventory of 4.3 million dollars[5]. Fashion giant Burberry has also destroyed finished products worth 38[6] million and this is increasing with time. This is an unsustainable madness. In such a situation, Nudges in terms of a 'buy-back' scheme has been helpful to tackle the problem of discarded clothes worldwide (Salaman et al., 2016). This scheme was introduced by the UK retailer John Lewis and later followed by other brands too. It's called "BuyBack with John Lewis," where the company buys unwanted and worn clothes from its customers and rewards them. These clothes are later recycled and reused for making new ones. This is an assertive step towards sustainability which also beautifully nudges the customers to choose this brand over others while appreciating its green initiative. Labeling products with psychological triggers are also an effective instance of nudging the consumers. Zalando[7] implemented this

---

[3] The Index of Industrial Production (IIP) is the index that exhibits the rates of growth in different industry groups of the economy in a specific time period.

[4] https://commerce.gov.in/trade-statistics/ (accessed on 12 December 2023).

[5] https://www.nytimes.com/2018/03/27/business/hm-clothes-stock-sales (accessed on 12 December 2023).

[6] https://www.bloombergquint.com/onweb/burberry-investors-question-destruction-of-38-million-in-goods (accessed on 12 December 2023).

[7] Zalando SE is a German multinational E-commerce company based in Berlin, Germany.

strategy by using various labels (such as the exclusivity or sustainability of the product) to capture individual attention. This is called the "Noble Edge Effect" (Chernev et al., 2015). Recent findings have shown verbal nudges appear to be more effective than visual nudges in terms of motivating consumers to buy sustainable apparels. It also increases the buyers' willingness to pay (WTP) for the sustainable alternatives (Roozen et al., 2021).

Indian fashion brands are also stepping into the field of nudge marketing. Raymond's recent ad campaign[8] – a series of four films display the facets of men's traditional wear such as bandhgalas, sherwanis, kurtas, etc. After the running of this campaign Raymond's has seen a significant surge in the sales of ethnic wear. Experts think the central message of the campaign, i.e., Celebrations are not unidimensional and can have multiple expressions no matter what the occasion is – this is what nudged the consumers towards purchase of ethnic formals. Myntra – an India based go-to marketplace for fashion products also nudges[9] its customers. Unlike other platforms when Myntra redirects you to a particular category of product a fashion related quote (usually by celebrities from the fashion industry such as designers, film stars, etc.) appears during the transition. Such quotes also appear at the end of the product details page. These quotes might appear insignificant to some, but they are very good strategies for nudging. Because when the customers constantly view such quotes associated with fashion, they subconsciously start to connect Myntra as the only reliable place for finding trendy and stylish apparels. Recommendations or Reviews are also important psychological triggers. Another India based fashion brand Nykaa uses this nudge effectively. Customer reviews are social proofs that help customers to make informed decisions. Nykaa provides incentives to its customers for writing online reviews (even the bad ones). It has customer reviews for almost all its products. This helps to reduce the customers' assessed risk and increases the probability of purchasing.

Nudge marketing is often confused with coercion. If the customers feel forced or persuaded to buy something – then it's definitely not a nudge (Elmqvist et al., 2021). However, we can identify some common parameters of nudge marketing especially for the fashion industry. These are the broad parameters observed and implemented successfully by some famous Indian brands also. As mentioned earlier, the term "Nudge" as a marketing tool is new to the Indian marketers, but they have been unknowingly using this tool

[8] https://www.exchange4media.com/advertising-news/grey-group-india-offers-a-fresh-take-on-customised-tailoring-in-raymond-campaign-100966.html (accessed on 12 December 2023).

[9] https://www.strategydive.com/post/how-myntra-uses-nudging (accessed on 12 December 2023).

successfully for quite a long time. However, the most common nudge marketing techniques observed in the Indian fashion market can be listed as follows: (i) product labeling; (ii) recommendations or reviews; (iii) price anchoring; (iv) decoy comparisons; and (v) exit intent overlays. Product labeling works as a psychological trigger by exhibiting the appealing features of the products to the customers or by emotionally motivating the customer by offering a limited period deal for an attractive product (Santos-Roldan et al., 2020). For example, silk mark[10] hangtags in silk sarees assure the customers about the authenticity of pure silk and emotionally nudge the customers to buy it in order to stop being fooled by cheap imitations available in the market. Recommendations by the brands for similar products are a clever way to nudge the consumer to buy similar things if the actual commodity is out of stock. This is primarily observed in e-commerce platforms. Customer reviews on the other hand, as mentioned before, are social proofs about the quality of the product and affect the consumer psyche immensely. Then comes Price Anchoring which displays a particular price compared to some other high prevailing-in-the-past benchmark price. We see this quite often especially during the festive seasons. Placing strikingly contrasting clothes or apparels next to each other is also a beautiful way to influence consumer buying decisions towards the one that is more attractive. This is known as Decoy Comparison. This strategy automatically makes one alternative relatively fancy or superior to the others and customers irrationally make their buying decisions without even realizing it. Exit Intent Popup is another effective strategy to influence consumer decisions in the fashion industry. This is applicable especially in the E-commerce platform. When a website visitor moves their mouse outside the boundary of the website, a pop-up message appears which shows offers that are more valuable than anything they have seen on the website (such as special discounts, coupon code, cash back offers, etc.). These five nudge marketing techniques along with other techniques and their effectiveness in the Indian context would be discussed in succeeding sections.

1.3 DATA AND METHODOLOGY

Our research is based on a primary survey. The motive is to identify the parameters that influence consumer decision making in the Indian fashion market. However, consumers are not familiar with the term "nudge," but they do know the factors which influence them directly or indirectly, i.e., the factors that they give weightage to while making fashion purchases.

[10] http://www.silkmarkindia.com/ (accessed on 12 December 2023).

Approximately 100 consumers were surveyed on a random basis. The respondents belong to various age groups but the majority of them are aged less than 20 years (53%). The study mainly analyzes the perception of the urban and semi-urban consumers as 93% of the total respondents are from urban and suburban areas. Only 7% are from rural areas. The questionnaire consisted of questions regarding the parameters influencing consumers while purchasing fashion products. It also had questions regarding the importance of the factors for a particular brand as the majority of the respondents identified themselves as brand conscious at some point of time. Hence, we divide our analysis into two parts – (i) examining the efficacy of nudge marketing parameters as a whole; and (ii) understanding how effective are brand-specific nudge marketing techniques in the Indian context.

The questionnaire used a five point Likert Scale[11] to understand the importance of the following factors in making fashion purchases – merchandise visibility, special discount on selected products, product description (e.g., ecofriendly material, recyclable, etc.), Customer Feedbacks/Reviews, Combo Offers (e.g., Buy 1 Get 1, etc.), Notification tags (e.g., Bestsellers, etc.), Subscription Benefits and Price Anchoring. All these are the standard nudge marketing techniques, as stated earlier. However, Special Discounts, Notification tags, Subscription benefits and Combo Offers are clubbed together and can be considered as a single factor, i.e., exit intent overlays. The second segment of the chapter focuses on effectiveness of brand-specific nudge marketing techniques. People, especially those from metropolitan areas, are usually brand conscious when it comes to fashion apparels. This motivates us to find out what are the most effective ways to nudge a consumer's preference towards a particular brand in the Indian context. Another five-point Likert scale is used to capture the same. The respondents were asked to rate the importance of the following factors for any particular brand – quality, price, number of physical stores, online presence of the brand, celebrity endorsement and product visibility.

1.4 FINDINGS

1.4.1 SUMMARY STATISTICS

From a sample size of 100, we have got some unique responses. Some of the responses we had to discard as they were outliers. After eliminating the

[11] A Likert scale is a psychometric scale commonly involved in research that employs questionnaires. It is the most widely used approach to scaling responses in survey research, such that the term is often used interchangeably with rating scale.

outliers, the number of effective responses were 97. The summary of our preliminary findings is provided in terms of Figures 1.1–1.7. Figure 1.1 shows the gender distribution of the respondents.

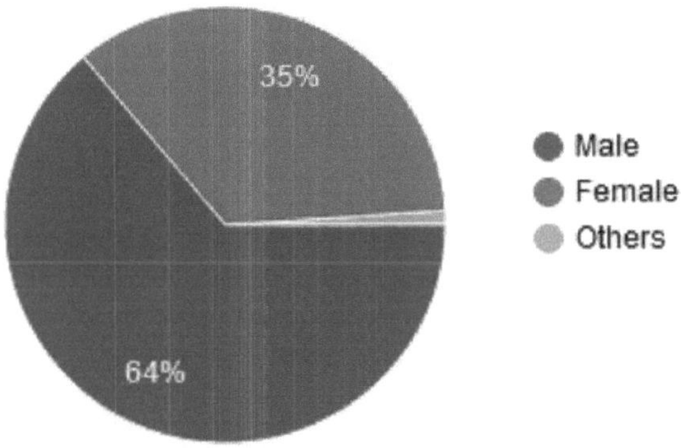

FIGURE 1.1 Gender distribution.

Figure 1.2 displays the age group of the respondents. And it has been observed that more than 50% of the sample size are teenagers. This finding is worth mentioning as teenagers are more trend conscious and brand loyal when it comes to fashion. Hence it matches our objectives perfectly.

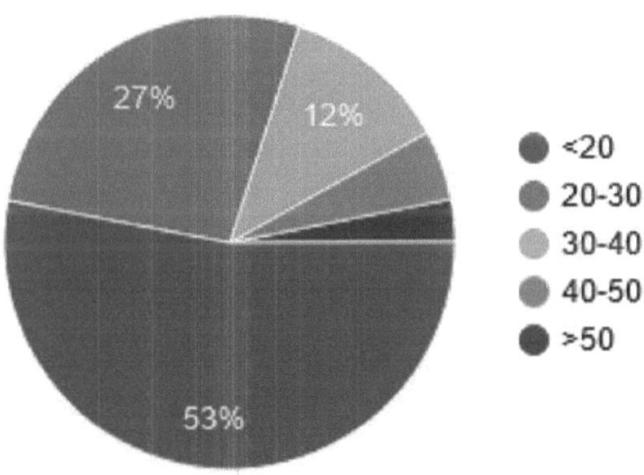

FIGURE 1.2 Age distribution.

The respondents mostly belonged from Urban and Semi-urban/sub-urban regions. This classification is depicted in Figure 1.3.

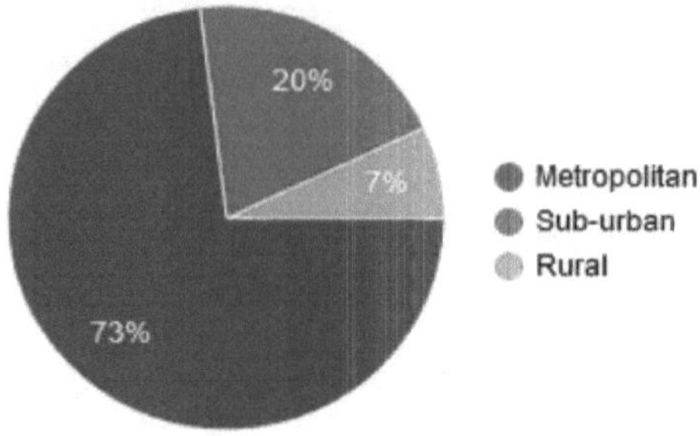

FIGURE 1.3 Regional distribution.

Figures 1.4 and 1.5 display two important aspects of our study. Figure 1.4 shows the amount of time spent (in hours) per week in checking out fashion products (both online and offline).

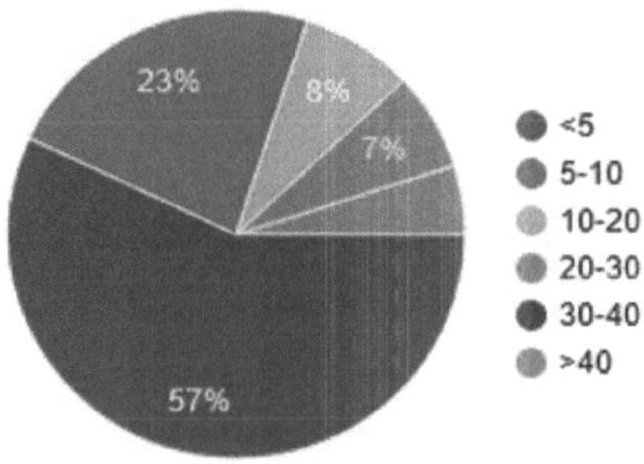

FIGURE 1.4 The amount of time spent (in hours) per week in checking out fashion products.

Figure 1.5 shows the frequency of purchasing fashion products in a month. In the regression analysis segment, the latter is considered to be the dependent variable.

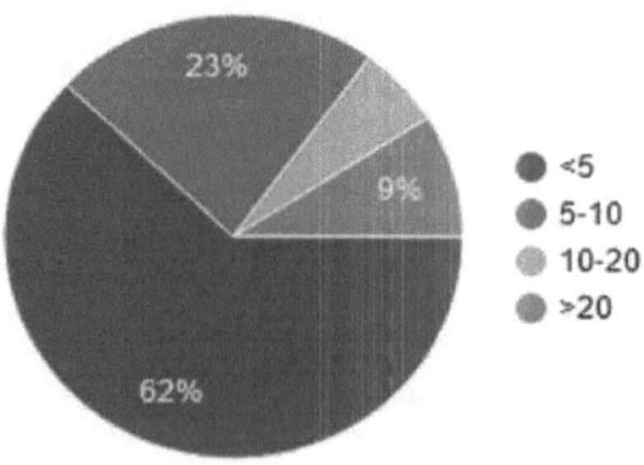

FIGURE 1.5 Frequency of purchasing fashion products in a month.

Figure 1.6 summarizes the monthly income parameter for all the respondents. We have specified six different ranges of income and as per the responses we could see that the majority of the respondents belonged to the monthly income category of less than 10,000 a month. This is perfectly consistent with the dominant age group and their designation as students.

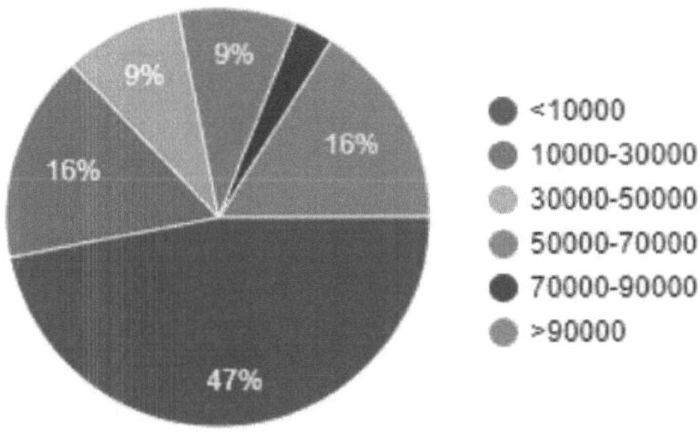

FIGURE 1.6 Monthly income.

Lastly, Figure 1.7 displays the respondents' brand consciousness of the respondents. This is another important aspect of our study. For the second objective this is used as the dependent variable.

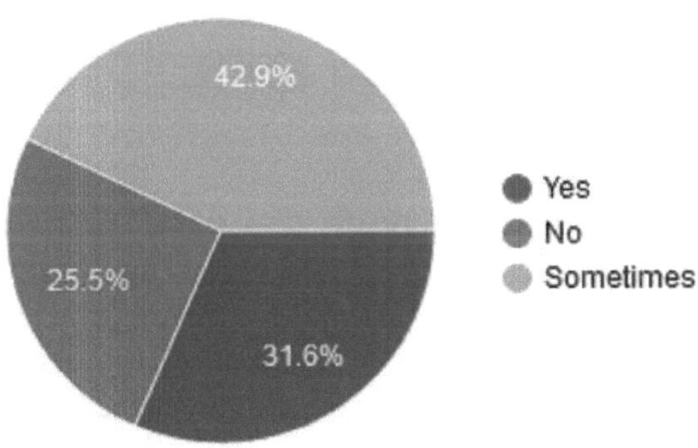

FIGURE 1.7 Brand consciousness.

1.4.2 REGRESSION RESULTS

As mentioned earlier, this chapter has two objectives. One is to examine the efficacy of nudge marketing techniques as a whole and the other one is to understand the effectiveness of brand specific nudge marketing techniques in the context of the Indian fashion Industry. For the first objective we have employed the ordinary least square (OLS) regression technique with 'frequency of monthly fashion purchase' as the dependent variable and five independent variables (Merchandise Visibility, Product Labeling, Feedback, Price Anchoring and Exit Intent Outlay). The responses of both the dependent as well as independent variables have been indexed after the elimination of outliers following the standard indexation technique. The formula used for calculating the indexes is as follows:

$$Index = \frac{Actualvalue - Minimumvalue}{Maximumvalue - Minimumvalue}$$

The result of the OLS regression is displayed in Table 1.1.

TABLE 1.1 OLS Regression Results

Frequency of Purchase	Coefficient	Std. Error	t	P>|t|	[95% Confidence	Interval]
Merch_visibility	0.024	0.16	0.15	0.087	–0.29	0.348
Product_labeling	0.014	0.202	0.12	0.902	0.426	0.376
Feedback	0.336	0.18	1.86	0.066	–0.693	0.223
Price_anchoring	0.034	0.239	0.14	0.078	–0.509	0.443
Exit_intent_outlay	0.504	0.339	1.47	0.144	–0.173	1.172
Constant	0.173	0.069	2.64	0.01	0.04	0.303

Source: Authors' calculation.

It is clear from the above result that the coefficients for Merchandise visibility, Customer feedback and Price anchoring are statistically significant at 10% level of significance. Thus, these three are the most important nudge marketing parameters that have significant impact on the monthly purchasing frequency of the consumers. The positive signs of the coefficients are also perfectly consistent with our hypotheses. The more the products are visible in the online and offline stores, the greater the positioning of the product in consumers' mind and the greater is the tendency to buy it. Again, if a product is reviewed (either word of mouth or written review or both) really well by the existing customers then there is a high possibility that other people will also have an inclination to buy it. Lastly Price anchoring can spark a subconscious response and encourage a purchase. It also happens to be an inexpensive and easily implementable strategy. Especially in the fashion industry this technique works wonderfully. Everyone likes to be up to date when it comes to fashion. Some of the common and successful anchor pricing techniques in the fashion field are strikethrough retail pricing, comparative package pricing, comparing prices with the competitors, etc.

VIF test was also performed along with OLS regression in order to check and correct for multicollinearity. The VIF results are as in Table 1.2.

TABLE 1.2 VIF Test Results

Variable	VIF	1/VIF
Merchandise_visibility	3.64	0.11
Product_labeling	8.23	0.19
Feedback	3.94	0.25
Price_anchoring	3.22	0.31
Exit_intent_outlay	8.48	0.37

Source: Authors' calculation.

As it is understood from Table 1.2, that two variables Merchandise visibility and Exit intent outlay exhibit serious correlations with other predictors of the model. Hence it is wise to drop them from the analysis and eventually after doing so it was found that the statistical significance of the three main predictors didn't change at all. The VIF result is also thus perfectly consistent with the main OLS regression output exhibiting high VIF (and low tolerance) for the two statistically insignificant predictors.

In order to fulfill the second objective, the logistic regression method has been employed. The dependent variable in this regard is 'Brand Consciousness' of the consumer. The respondents were asked whether they identify themselves as brand conscious individuals or not. The answers would be either a "yes" or "no" and the responses have been coded as 1 and 0, respectively. Hence it is a dichotomous variable and that gives the justification of employing logistic regression. The chosen independent variables for this are quality, price, number of physical stores, online presence of the brand, celebrity endorsement and product visibility. The indexation of the responses for the independent variables are also done in the way mentioned earlier. The logistic regression output is shown in Table 1.3.

TABLE 1.3 Logistic Regression Results

Brand Consciousness	Coefficient	Std. Error	z	P > z	[95% Confidence	Interval]
Quality	3.077	1.451	2.12	0.034	0.233	5.921
Price	–3.480	1.495	–2.33	0.020	–6.409	–0.551
Physical stores	–0.866	0.981	–0.88	0.377	–2.787	1.056
Online presence	1.372	1.118	–1.23	0.022	–3.563	1.819
Celebrity endorsement	2.095	1.037	2.02	0.043	0.063	4.127
Product visibility	0.989	1.231	0.80	0.422	–1.423	3.400
Latest trend	0.409	1.141	0.36	0.720	–1.828	2.646
Responsibility of the brand	–0.956	1.204	–0.79	0.028	–3.316	1.405
Constant	0.950	0.599	1.59	0.113	–0.224	2.124

Source: Authors' calculation.

When it comes to brand specific nudge marketing techniques, quality, price, online presence, celebrity endorsement and responsibility demonstrated by the brands happen to be the most significant parameters – all at 5% and 10% level of significance. The signs of the coefficients are also perfectly consistent with the hypotheses. The better the quality of the product offered by the brand, the more brand conscious or brand loyal the customer will become. Also, the price charged by the brand negatively impacts brand consciousness. If the price increases exorbitantly, then the brand will lose some of its customers. On the other hand, given the present scenario it is of no doubt that Online presence of the brands will attract more customers. Along with that the Celebrity Endorsement parameter is also significant. The primary reason behind this is the demographic composition of our data. As more than 50% of the respondents are teenagers, it is quite understandable that if their favorite movie star or sports person endorses a particular brand, they will think that particular brand is 'cooler' than others and will stick to that brand for a long time. Last but not the least, the fact that a brand is responsible towards the environment and/or the society also nudges the customers in a positive direction. Much evidence can be found in this regard. Especially the brands promoting 'sustainable fashion'[12] have received many positive responses. The customers associating themselves with these brands feel that they are a part of something bigger and better

[12] Sustainable fashion is an all-inclusive term describing products, processes, activities, and actors aiming to achieve a carbon-neutral fashion industry, built on equality, social justice, animal welfare, and ecological integrity.

for the society and hence they are most likely to remain loyal with the brands.

1.5 CONCLUSION

Nudge marketing is a combination of all such marketing techniques that follow 'Libertarian Paternalism.' 'Libertarian' means the consumers are 'free to choose' and 'Paternalism' means guiding people in a certain way or to choose a particular product. The nudge marketing parameters chosen in our study follow this principle and rely on cognitive biases. In the field of nudge marketing, a consumer is never forced to choose something in favor of the producer, rather it influences the consumer psyche and drives them in a particular way without letting them realize. This is a preliminary study based on the effectiveness of behavioral nudges on the consumers of the fashion industry, which can be further extended to cover other areas as well and see how the results change. This research paves the way to consider the efficacy of nudge marketing techniques of other Indian industries as well. It will assist in strategy framing to tackle this kind of response generating out of herd mentality to maximize its effect on the mass and the economy as a whole. This study also has some limitations. The comparisons of nudge marketing techniques in the online and offline mode have been considered together. So, whether online efficacy is more than the offline efficacy – we are unclear about that. Nonetheless, this opens up scope for future research in this particular field.

The Indian fashion industry is huge, dynamic, and ever-changing. We are living in an age where the consumers want transparency and social responsibility. Given these, the traditional marketing techniques may not be effective in some cases. Alternatively, nudge marketing techniques are likely to be more successful where traditional marketing fails as it relies on consumers' cognitive biases. Also, there is enough evidence that the term 'nudge' marketing might be unknown to the Indian marketers, but they have been using such marketing techniques successfully for quite a while, especially in the fashion industry. Hence, with a change in the demand of the markets and changing the psyche of the consumer, it is high time to adopt newer marketing strategies and implement them successfully so that it turns out to be a win-win situation for both the consumers and producers. Nudge marketing techniques perfectly assure that.

KEYWORDS

- **behavioral economics**
- **brand consciousness**
- **consumer behavior**
- **fashion industry**
- **libertarian paternalism**
- **logistic regression**
- **nudge marketing**

REFERENCES

Ahuja, N., (2015). Effect of branding on consumer buying behavior: A study in relation to fashion industry. *International Journal of Research in Humanities & Social Sciences*, *3*(2), 32–37.

Chernev, A., Böckenholt, U., & Goodman, J., (2015). Choice overload: A conceptual review and meta-analysis. *Journal of Consumer Psychology*, *25*(2), 333–358.

Cheung, B., Letherby, A., & Pearce, A. (2021). Nudge marketing. *L'Économie*, *80*.

Elmqvist, J., & Thorell, J., (2015). *Nudge Marketing: How to Influence Decisions by Changing the Choice Architecture.*

Goldstein, D. G., Johnson, E. J., Herrmann, A., & Heitmann, M., (2008). Nudge your customers toward better choices. *Harvard Business Review*, *86*(12), 99–105.

Gupta, M. I. (2018). A study on use of nudge marketing by Indian car marketers, making neuro marketing a practical reality. *Financial Management & Marketing Management*, 492.

Roozen, I., Raedts, M., & Meijburg, L., (2021). Do verbal and visual nudges influence consumers' choice for sustainable fashion? *Journal of Global Fashion Marketing*, *12*(4), 327–342.

Salaman, G., & Storey, J., (2016). *A Better Way of Doing Business?: Lessons from the John Lewis Partnership*. Oxford University Press.

Santos-Roldán, L., Palacios-Florencio, B., & Berbel-Pineda, J. M., (2020). The textile products labelling analysis and requirements. *Fashion and Textiles*, *7*(1), 1–24.

Singh, S., (2019). Nudging behavior in marketing dynamics: Behavior economics and marketing-an interlinking. *Drishtikon: A Management Journal*, *10*(2), 39.

Thaler, R. H., & Sunstein, C. R., (2003). Libertarian paternalism. *American Economic Review*, *93*(2), 175–179.

Thaler, R. H., (2017). Behavioral economics. *Journal of Political Economy*, *125*(6), 1799–1805.

Tversky, A., & Kahneman, D., (1989). Rational choice and the framing of decisions. In: *Multiple Criteria Decision Making and Risk Analysis Using Microcomputers* (pp. 81–126). Springer, Berlin, Heidelberg.

CHAPTER 2

Commodity-to-Brand Transformation: A Review and Research Agenda

DIBYENDU NANDY[1] and SUDIPTA MAJUMDAR[2]

[1]*PhD Scholar, Faculty of Management Studies, ICFAI University, Jharkhand, India*

[2]*Associate Professor, Department of Management, School of Business & Economics, Adamas University, India*

ABSTRACT

The purpose of this chapter seeks to develop a framework for understanding what a commodity is and what factors drive the transformation of a commodity to a brand. The design/methodology/approach of this chapter reviews empirical studies made in the transformation of a commodity to a brand. It compares and analyzes the results from exploratory field studies and conceptual articles with prior research on the transformation of a commodity to a brand. The findings of this study find that product, service, and image are prime independent constructs in transforming a commodity to a brand with the involvement of both the producer and customer. Creation of value is critical to creating a brand with relevant variables like quality of product, packaging, accessibility, responsiveness, trust, emotional, and social value contributing to the transformation. Value generation is identified as a dependent construct and its nature and extent of linkage with independent constructs have been proposed for future research for validation and further investigation. The practical implications of this chapter are the development of a conceptual model that brings to the table a multi-dimensional structure for independent constructs and the dependent construct, individually as well as collectively, for contributing towards the transformation of a commodity to a brand. The

Sustainability in Marketing Practice: Strategies for Industry 4.0.
Subrata Chattopadhyay, Udit Chawla, (Eds.)

originality/value of this chapter provides a framework for understanding, conceptualizing, and identifying drivers of transformation. The chapter puts forth research questions, which on further exploration would provide insights to marketers to decide on marketing spend and customized value proposition across customer segments. Core sectors like coal, cement, agriculture, energy, metals, sugar, etc., mainly form the commodity market. Customers, if they fail to observe a difference between products, tend to label it as commodities. A product without any individuality becomes a commodity and in such a state, consumers do not show any specific preference for an item. Brands play a vital role in delivering identity and creating differentiation. The chapter takes a critical review of the studies on commodity branding conducted in the past, constructs a conceptual framework, provides an understanding of the branding process, and examines the various dimensions of branding elements which can be used effectively to create a multi-dimensional construct.

2.1 INTRODUCTION

Commodity markets are driven by bulk transactions of products such as sugar, cotton, metal, energy which all are undifferentiated can be easily replaced with comparable products. A product without any identity turns to a commodity and in that case, consumers do not show any liking for an item. In a competitive market scenario, marketers try to craft that identity to get noticed and perceived differently from competition. Reality is even if actual differences exist, if customers do not perceive it then it does not mean anything. Marketers have determinedly tried to establish differentiation either through superior product characteristics or through improved serviceability. In a homogeneous market, branding of commodities takes place when customers perceive a high value for the product and the product is appealing and creates excitement for them. The focus of commodity branding is to create significant differences through involvement of customer involvement and engagement to move from commodities to a customer-developed brand and further to a strong corporate brand. Commodity branding benefits consumers as well as marketers. The consumers gaining more confidence and trust on the product by mitigating the risk, marketers get benefited by attracting premium for their products which the customers are eager to pay. Normally, commodities are primary products that undergo slight processing and then gets sold in the market. Commodities are normally divided into two categories: hard and soft commodities. Soft commodities are those which are perishable in nature, like coffee, sugar, cotton, etc., while hard are those which can be stored for longer periods. These include oil, natural gas, and precious metals. In a commodity market, customers buy equivalent products from different producers. The undifferentiated characteristics of commodities leads to restricted customer faithfulness to any producer as customers mostly purchase based on pricing. This effectively leads to severe competition among producers resulting in low profit margins. Thus, producers competing in a commodity market try to create differentiation by value addition to the product with packaging, delivery, and several other parameters. To create differentiation and stand out and to give customers a compelling reason to buy, commodity producers started attempting to brand their products. Branding is not new, as per Archaeologist David Wengrow, branding started around 5,000 years ago in the city of Uruk, now Iraq, villagers made stone seals to mark caps and stoppers of food and beverages. The villagers trusted the stone seals on caps and stoppers of oil and wine as dependable and reliable. Wine is one of the first products considered branded as the containers were marked with marks by the manufacturers to distinguish from others (Duguid, 2003; Levitt, 1980) brings out the fact that in 1900s in Soviet Union generic televisions were manufactured from several companies, and with quality being inconsistent, customers used to look for product codes at the bottom of the television to discover the source of manufacturing and buy the product of their choice. The study of Michel (2008) reveals that unless a buyer perceives a difference between products, they are not able to differentiate on their own as perceived difference is only reality. With similar characteristics and attributes and with a wide range of products where buyers do not perceive differences, commodities create confusion in the minds of the buyers, Brands bring order to this confusion and chaos by creating perceived differences in the minds of the customers. A buyer when confronted with confusion wants a solution which would mitigate the risk of dissonance and provide the best economic value. Brands help mitigating this risk and provide value which is higher than the price the buyer is willing to pay. Higher the economic value better for the buyer. Now the question comes "What do the Brands do?" Brands do the following:

- Provide identity;
- Differentiate;
- Mitigate risk;
- Facilitate relationships;
- Create value; and
- Build culture.

While commodities lack identity as they are similar in appearance and characteristics, marketers and consumers have tried to create identity for the commodities to bring order to chaos and ensure the selection of the right product. Pennington et al. (2009) illustrate with an example of creating the identity of bread in New York City in the late 1800s, when people bought bread from street vendors who used to procure from bakers. Bakers used to specify their loaves with a mark representing full weight to demarcate from short-weighed loaves. Each baker used to have their own marks; however, customers quickly learned to distinguish full-weight loaves from short-weighed loaves with the marks and chose to buy loaves with full weight only. Identification and differentiation, being the primary functions of the brand, could provide distinctiveness to customers and at the same time also help in providing assurance, mitigating risk, and making the right choice. Customers can also become emotionally attached to a brand over a period when they see their self-image in the brand and the implicit promise is delivered. The relationship with the brand leads to customer attachment and loyalty. Marc Gobe (2010) in his book "Emotional Branding: The New Paradigm for Connecting Brands to People" opines that personalization of a product results in a higher chance of bonding with a brand. The emotional attachment and loyalty ensure that customers stick to the brand and find no reason to switch to a competing product. In a sense, brands create both emotional and economic value for the customers. It increases the value of the business and becomes the company's most valuable asset. Rightly stated by Stephen King of WPP, "A product is something that is made in a factory; a brand is something that is bought by a customer." It is important to communicate to customers to tell the story of the brand which will captivate and convince customers and put forward intrinsic or core values of the brand to the customer for which it stands for. Brand culture reflects the company, and all marketing actions should be coordinated to reflect the image and values that the brand inculcates in the minds of the consumer. Now the question comes "Can Commodities Be Branded?" (Levitt, 1980) answers this question straightforwardly when he clearly says that all goods and services can be differentiated. Several approaches have been taken for branding of commodity products, and marketers and customers have tried to create differentiation on different parameters. These include - (i) product attributes and marks; (ii) quality or standards; (iii) performance with respect to customer expectations; (iv) distinctive origin; (v) service; (vi) perceived image; and (vii) customer involvement and engagement. In this chapter, we are trying to review the literature to find out the prospective research gaps and the scope of research in the domain of the transformation of a commodity.

2.2 LITERATURE REVIEW

In this chapter, we intend to examine the approaches to brand commodities. The two main approaches as could be identified through study of several literatures are:

- Customer branding of commodity products; and
- Manufacturer branding of commodity products.

2.2.1 CUSTOMER BRANDING OF COMMODITY PRODUCTS

The landmark study by Pennington et al. (2009) on customer branding views it as a "procedure by which a customer on their own hunt for parameters to be able to create differentiation on their own." Aaker (1997) and numerous authors have illustrated how to build brands by manufacturers and service firms, while on the contrary, Duguid (2003) postulated the reverse order where customers being active to communicate their perspectives of branding to producers. This emphasizes the importance of customer branding wherein the customer proactively creates differentiation on his own. The classic case of Vidalia Onions is an example where the customers tried to differentiate by identifying the source of origin. As narrated by Clemens (2005), the story of Vidalia Onions when Georgia farmers started marketing sweet onions in the local market of Vidalia. The travelers who consumed the sweet onion demanded more of the same, and local farmers started supplying even from regions which are not within the peripheries of Vidalia. The continuous communication from customers eventually made the onion a national brand, and in 1986, the State of Georgia registered the trademark "Vidalia," and currently Vidalia Onions are marketed worldwide from Georgia. Customers turned the commodity into a brand on their own by giving them a separate identity. Customer branding is different from normal branding, as in normal branding, the communication flow is from marketer to customer while it is reversed in the case of customer branding. Turning a commodity into a brand, however, has benefits for both marketer and customer as identified is depicted in Table 2.1.

TABLE 2.1 Benefits of Branding

Benefits of Branding for	
Marketer	**Customer**
Identifies the company's product, build image.	Helps to mitigate perceived risk in buying.
Allows to attract premium to get better margins.	Provides assurance of reliability, quality.
Allows to develop the brand for new products, new market, and new geographical areas.	Offer psychological reward.

Source: Adapted from Jha (2014)

The three conditions of customer branding to happen came out from the study of Pennington et al. (2009), which specifies that product variation, delivery of general product must be improper, and the customer must depend upon a suitable subgroup which satisfies customer needs. In customer branding, the customer interprets the market as heterogeneous, while the marketer finds it homogeneous. The product variation in the marketplace must exist for customers and the product offered is not able to meet customer needs, the customer tries to find a differentiator in the supply chain which distinguishes the product for the customer with preferred attributes. Clemens (2005) brings out the typical case of I-80 beef in which Japanese retailers were asking for Iowa-80 beef processed in Iowa and Illinois. The retailers particularly asked for I-80 beef which is their own creation and in turn excluded beef from other regions which had comparable properties. Iowa's reputation has been built on the state's produce of well-marbled beef with tenderness. These traits are especially important in niche markets of Japanese consumers where consumers value well-marbled, tender beef. Evolution of customer branding has taken place over the years and Pitt et al. (2006) illustrates the evolution of customer brand model from commodities to an open-source brand (Figure 2.1).

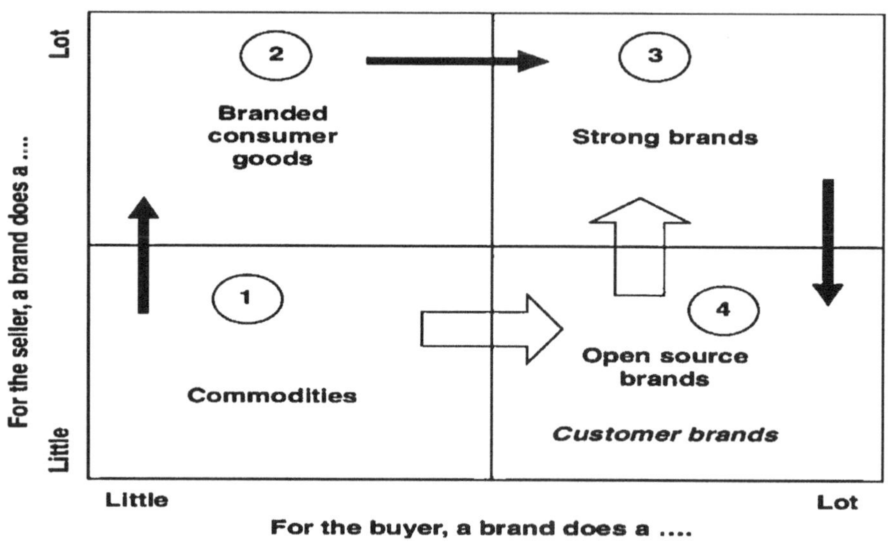

FIGURE 2.1 Evolution of brand.

Source: Reprinted with permission from Pitt et al. (2006). Copyright Springer Nature.

The evolution of brands starts from Commodities to Branded Goods to strong corporate brands and finally to open-source brands. In the first stage are the Commodities, where there is little scope for the buyer or for the seller as products are undifferentiated. In the next stage, marketers can segment customers, go for promotion, and try to realize premium, whereas customers do little. The third stage is the stage where the customer does a more informative search to mitigate risks, while the marketer too makes all attempts to convince, motivate, and simplify the decision-making process of buyers. The last stage of the model advocates conversion to Open-Source Brands where customers or buyers are actively engaged and there is not much scope for marketers in terms of branding. For Open-source brands, customers become participants in co-creating the brand, suggesting differentiation, labeling, and communication, but in the process, the brand may lose its novel architecture and become more generic. Pennington et al. (2009) cites examples of the Linux Operating system which has undergone several transformations through the hands of customers in which they have customized it to suit their requirements. However, some famous brands like P&G have developed a counter-clockwise trajectory as cited by Pitt et al. (2006) to the fact that some famous brands have not followed the path as established by Pitt et al. Customer Branding is a phenomenon which has been explored to a certain extent in marketing literature and understanding the circumstances when and how customers engage themselves to brand commodities paves the way for marketers to know customer needs better and thereby helps them to identify areas of differentiation and improvement. However, Customer Branding is an area which requires further investigation to fully comprehend the phenomenon, especially in which parts of the value chain customers take part and how.

2.2.2 MANUFACTURER BRANDING OF COMMODITY PRODUCTS

Unger (1983) classifies commodities based on the criteria of undifferentiated/differentiated and high/low volume (Table 2.2).

TABLE 2.2 Classification of Commodities

	Undifferentiated Products	Differentiated Products
	True Commodities	**Pseudo Commodities**
High Volume	• Sold at relatively low unit values.	• Produced to accepted performance specifications but with minor differences.
	• Widely used in a variety of applications by many customers.	• Sales concentrated in few large customers.
	• Sales concentrated in a few large customers.	• Some degree of differentiation exists.
	• Contract pricing	–

TABLE 2.2 *(Continued)*

	Undifferentiated Products	Differentiated Products
	True Commodities	**Pseudo Commodities**
	Fine Products	**Specialty Products**
Low Volume	• Substantially identical product form and composition.	• Differentiation by formulation.
	• High unit prices.	• Produced by various suppliers based on performance in use.
	• Small numbers of customers in low or moderate volume.	• Designed to solve customer's specific problems.
	–	• Relatively high price.
	–	• Large number of customers.

Source: Reprinted with permission from Robinson et al., (2010). Copyright Taylor & Francis.

Though Unger and Emerson developed the grid in classifying commodity chemicals, the same can be logically extended to other commodity products (Robinson et al., 2002). Over the years, marketers are progressively adapting branding strategies to convert commodities to brands to respond to increasing competition. To distinguish one commodity product from another, marketers have made branding efforts which combine product differentiation, supply chain, service, and creating an image in the mind space of the consumers. The very nature of commodities makes them susceptible to tremendous price pressures which erode the margins of the producers. To safeguard their margins, commodity product manufacturers felt the need to create a competitive sustainable advantage by building a long-term loyal customer base. Branding is the approach adopted by marketers to persuade customers to buy the desired product to satisfy the need. As stated by Docherty (2012), branding is not just limited to glossy advertising, it encompasses everything that distinguishes a product or service. With reference to a study by Punniyamoorthy (2011), branding is a construct which is multidimensional in nature and the creation of value and delivery of the same are the cornerstones of any successful commodity branding. Studies in the past by Chaudhuri & Holbrook (2001), Maity (2017), Mudambi et al. (1997) have indicated that branding of commodities, being a multidimensional construct, is not limited to how customers view the physical product but the service and image that accompanies the product. This points to creating differentiation – either through Product or through Service or through creating Imagery. Commodity product marketers are more settling for branding to respond to competition. Brands help build relationships with customers which, in turn, builds loyalty. Loyalty is a barrier to brand switching and ensures repeat purchase. Commodity branding is a technique used by marketers to create a competitive edge by utilizing the brand so that they can be perceived differently from others.

2.3 RESEARCH GAPS

In a commodity market where there is a lot of homogeneity and a quagmire of products, consumers struggle to choose a particular product; brands provide a means to sort out the confusion. The decision of buying for a consumer in a commodity market, which hinges upon price factors, changes to attainment of value out of using the product as brands support them with reasons to go for a product which they feel true to meet their expectations. The research gaps that could be identified in literature review can be summarized as:

- Is emotional/aesthetical value addition important for commodity to brand conversion?
- How does labeling and Packaging and naming strategies help to differentiate a commodity from a brand?
- Do associations and endorsement of celebrities help in transforming a commodity to a brand?
- Is customer involvement, price worthiness, social value makes any significant impact on transformation of commodity to brand?

2.4 CONSTRUCTS

Numerous studies on commodity branding have been conducted in the past to unearth the relevant strategies of transforming a commodity to a brand. Three main constructs could be identified viz Product, Service, and Image on which the conceptual framework is proposed later. Product differentiation, as studies done in the past, reveal quality (Vertinsky & Zhou, 1999), packaging and labeling (Norberg et al., 2011), features (Romaniuk & Sharp, 2004), name and logo as main components on which marketers have tried to create an identity. Wengrow (2008) illustrates labeling of mass-produced goods aids in quality control, originality, and ownership which contributes to build the modern brand. The construct of Product can be diagrammatically represented in Figure 2.2.

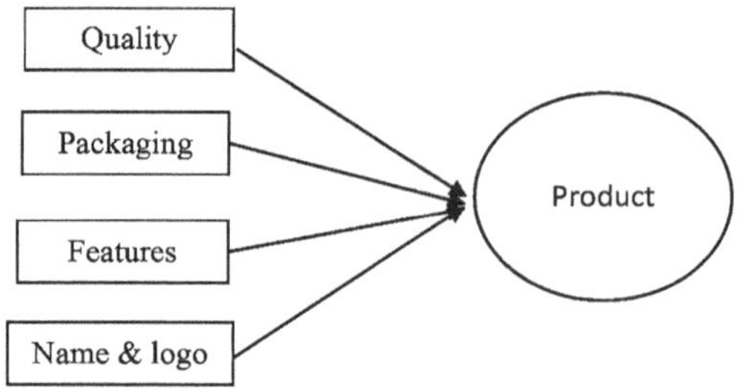

FIGURE 2.2 Construct: Product.

It is not only about the tangible factors which create differentiation, but also intangible factors matter even in rational decision making as found in studies by Robinson et al. (2002); Mudambi et al. (1997); Berry (2000); and Shostack (1989). The study on electricity sector by (Larsen, n.d.) brings out the fact that the service part of distributing energy is important. In addition, the study also indicated the need to differentiate based on service as the product is undifferentiated, the only difference that can be created is in providing the service with the product. The study on commodity chemicals by Robinson et al. (2002) clearly emphasizes the need for a differentiation strategy to come out from cost leadership which most commodity makers would make use of to compete in the market. Prior studies by Grönroos (1993) and others identified the following parameters like accessibility (regular contact with customers), delivery on time, order handling procedures, responsiveness (reaction time to queries and complaints), and Information flow (Clarity in communication) function as differentiators in an otherwise undifferentiated product category of commodities as given in Figure 2.3.

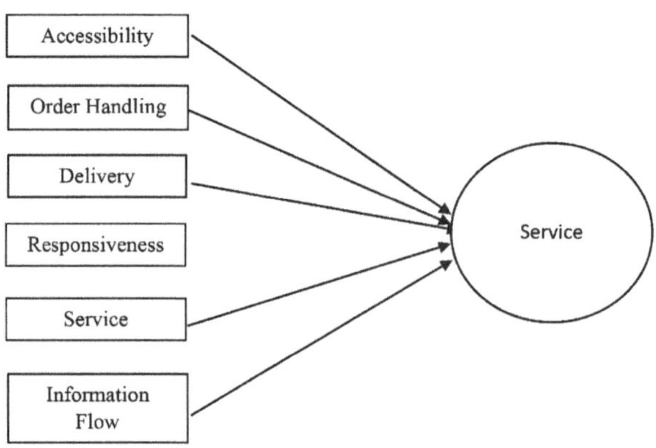

FIGURE 2.3 Construct: Service.

The pinwheel of added value to the customer in terms of intangible aspects shows a way of transforming the commodity to a brand; however, it is equally important for the marketer to effectively communicate the value created for the customer. The communication of the value would play a pivotal role in creating a favorable image in the minds of the customer. The purpose in creating an image is to communicate to customers a promise in terms of safety, by removing the perceived risk associated with purchasing commodity products. Brand strategies are principally meant for creating unique brand images. The aim of branding is to build awareness and knowledge among customers to build loyalty and induce customers to buy and make repeat purchases. Some studies in the past aimed to understand whether the same works in a commodity category with respect to perceived differentiation created by marketing actions. Grönroos (1993) opines that brand image can be created by perceived differentiation apart from product and service differentiation. Larsen (2010) is of the opinion that perceived differentiation has a motivational perspective (Aaker, 1997) and would provide buyers with a reason to purchase and remain loyal to the brand. Morton & Zettelmeyer (2004) state that brand loyalty has been a primary focus of marketing planning and serves as an important parameter for developing a sustainable competitive advantage that can be achieved through marketing actions. Chaudhuri & Holbrook (2001) brand loyalty model springs from research conducted on surveys of consumers. The antecedents of brand loyalty as identified are brand trust, purchase loyalty, attitudinal loyalty, and relative price. Further studies by Giddens (2015) indicate awareness, perceived relevance, and perceived value as important parameters for brand loyalty and creating a brand image. Uncles (2010) have two main propositions for loyalty - one is more qualitative by building a closer bond with customers and the other more quantitative as an increase in sales revenue by increasing usage levels. Nl et al. (2016) in their study indicated overall customer satisfaction, trust, switching cost, and reputation as determinants of brand loyalty. Adeyemi et al. (2013) in their study on the beauty and healthcare sector propose a multidimensional model using a chaining process and a moderating variable. The attributes that the study reveals are performance, perceived value, trust, behavioral, and attitudinal loyalty. The study of Punniyamoorthy (2011) in the newspaper industry describes brand loyalty as a construct which includes attitudinal commitment as well as behavioral purchase loyalty. While the majority of studies focus on measuring brand loyalty of branded products, Punniyamoorthy (2011) focused on a study to measure brand loyalty score for commodities with a reference to the cement industry. The AHP (analytical hierarchy process) model used in the study embraces qualitative and quantitative aspects of brand loyalty and tries to build it as a comprehensive measure. The suggested methodology is a step towards mitigating risks to build a loyal customer base for commodities. Independent variables as identified in the study are Involvement, Price worthiness, social value, emotional/esthetic value, satisfaction, trust, and the dependent value is perceived brand image and loyalty as depicted in Figure 2.4.

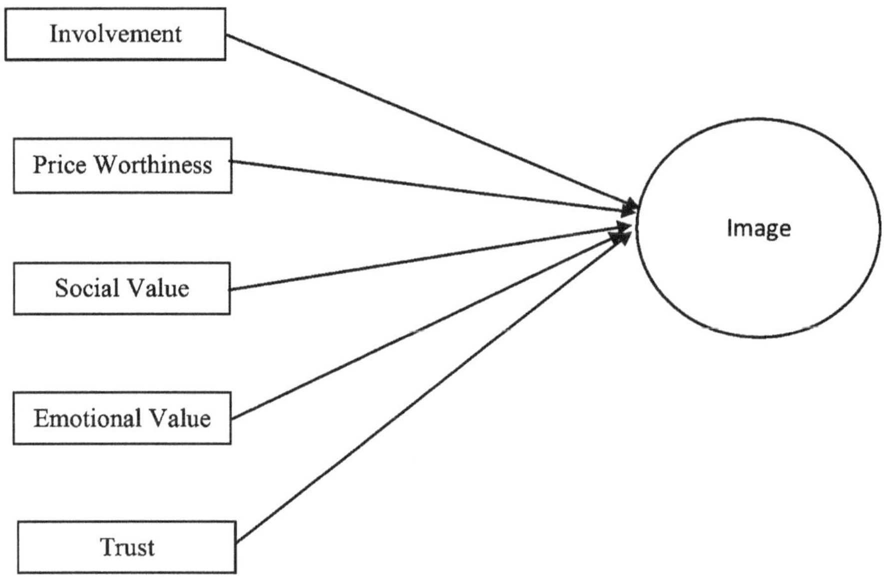

FIGURE 2.4 Construct: Image.

2.5 CONCEPTUAL FRAMEWORK

The proposed conceptual framework combines the three constructs holistically to have a comprehensive view of the constructs contributing to transformation of commodity to brand and apply the model in the future to test the extent of each construct in contribution to transformation as given in Figure 2.5.

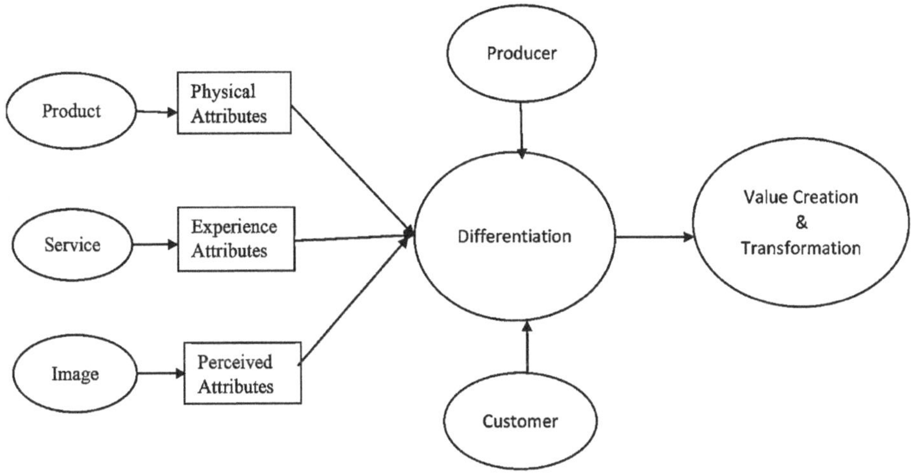

FIGURE 2.5 Conceptual framework.

The three constructs as elucidated above are Product, Service, and Image. The product construct talks about the physical attributes which includes quality, features, logo, name rendering an identity to the commodity, while the service construct relates to experience attributes which a customer goes through information search, delivery, accessibility, and responsiveness. The third construct, i.e., image, as perceived by customers, is related to price worthiness, social and aesthetic value, trust. All the three constructs contribute and combine to form differentiation both from customer and producer perspectives. Customer along with producer can also co-create the differentiation in a combination of either or all the three constructs to create value and thereby transform the commodity to brand. The transformation of commodity to brand does not happen either by only Customer Branding of Commodity products or by only Producer Branding. The customer gets proactively involved in creating differentiation in the value chain when the need to differentiate is felt to overcome the confusion dealing with homogeneous products. Manufacturers of commodity products have taken the route to distinguish their products from others through use of various marketing tools – both tangible and intangible. For example, Pilgrim's Pride Egg Plus, a separate brand of eggs with Omega-3 fatty acids and Vitamin E is different from normal eggs which is a commodity product. Bricks though being a commodity, Acme Brick created perceived differences through advertising campaign to educate house builders on how to recognize Acme Bricks carrying a different seal and assuring a 100-year performance guarantee.

Thus, both real and perceived differences are important in transformation of commodity to brand.

2.6 CONCLUSIONS AND SCOPE FOR FUTURE RESEARCH

The studies done in the past are not exhaustive and have been limited to either specific sectors or specific sections of customers. Moreover, using the framework, the areas or gaps identified can be further studied to explore both tangible and intangible aspects and measure their contributions in transforming a commodity to a brand. The contributions of each construct in the transformation of a commodity to a brand, if measured, can help marketers optimize their marketing spend and design a more impactful offering to the targeted groups.

KEYWORDS

- **analytical hierarchy process**
- **brand transformation**
- **branding**
- **commodity**
- **customer branding**
- **differentiation**
- **omega-3 fatty acids**
- **value**

REFERENCES

Aaker, J. L., (1997). Dimensions of brand personality. In: *Source: Journal of Marketing Research* (Vol. 34, No. 3). http://www.jstor.org; URL: http://www.jstor.org/stable/3151897 (accessed on 8 December 2023).

Adeyemi, B. O., Aiyenigba, E., Ademola, A. O., Makanjuola, A. T., Ibrahim, M. O., Ganiyu, S. A., Oluwakemi, B. O., & Ademola, A. O., (2013). Primary health care consumers' perception of quality of care and its determinants in North-Central Nigeria. In: *Journal of Asian Scientific Research* (No. 3). https://www.researchgate.net/

publication/259467561_Primary_Health_Care_Consumers%27_Perception_of_Quality_of_Care_and_Its_Determinants_in_North-Central_Nigeria (accessed on 8 December 2023).

Berry, L. L., (2000). Cultivating service brand equity. In: *Academy of Marketing Science Journal; Winter* (Vol. 28).

Biel, A. L., (n.d.). *Discovering Brand Magic: The Hardness of the Softer Side of Branding.* https://doi.org/10.1080/02650487.1997.11104688 (accessed on 3 January 2024).

Chaudhuri, A., & Holbrook, M. B., (2001). The role of brand loyalty. The chain of effects from brand trust and brand affect to brand performance: The role of brand loyalty. In: *Journal of Marketing* (Vol. 65).

Clemens, R. L. B., (2005). Geographical indications, the WTO, and iowa-80 beef. In: *Iowa Ag Review* (Vol. 11, No. 2).

Docherty, C., (2012). *Branding Agricultural Commodities: The Development Case for Adding Value Through Branding.* In International Institute for Environment and Development. https://www.iied.org/sites/default/files/pdfs/migrate/16509IIED.pdf (accessed on 8 December 2023).

Giddens, N., (2015). Building your brand. In: *Ag Decision Maker Newsletter* (Vol. 6, No. 3). http://lib.dr.iastate.edu/agdm/vol6/iss3/3 (accessed on 8 December 2023).

Grönroos, C., (1993). A service quality model and its marketing implications article. *European Journal of Marketing.* https://doi.org/10.1108/EUM0000000004784 (accessed on 8 December 2023).

Jha, M. (2014), Effective Marketing Strategy For Branding A CommodityWith Special Reference To Consumer Goods, International Journal of Social Science & Interdisciplinary Research, ISSN 2277-3630

Larsen, F., (2010). *Review and Conceptualization of Commodity Branding.* University of Economics in Katowice.

Larsen, F., (n.d.). *Branding As a Bridge for Commodities Towards a Liberalized Market.* https://cejsh.icm.edu.pl/cejsh/element/bwmeta1.element.desklight-7a4a6560-de68-4d65-a3f2-4673bc027348 (accessed on 8 December 2023).

Levitt, T., (1980). *Marketing Success Through* (pp. 83–92). Harvard Business Review.

Maity, A., (2017). A comparative study of branding strategies harnessed by Indian cement manufacturers. *IUP Journal of Brand Management, 14*(2), 64–69. http://search.ebscohost.com/login.aspx?direct=true&AuthType=ip,shib,cookie,url&db=bth&AN=123727547&site=ehost-live (accessed on 8 December 2023).

Morton, F. S., & Zettelmeyer, F., (2004). The strategic positioning of store brands in retailer-manufacturer negotiations. In: *Review of Industrial Organization* (Vol. 24).

Mudambi, S. M., Wong, V., Mcdowell, S., Doyle, M. P., & Mudambi, S., (1997). *An Exploration of Branding in Industrial Markets.* https://doi.org/10.1016/S0019-8501(96)00151-4.

Nl, W. O., Schijns, J. M. C., & Caniels, M. C. J., (2016). The impact of perceived service quality on customer loyalty in sports clubs. *International Journal of Sport Management Recreation and Tourism, 24*(C), 43–75. https://doi.org/10.5199/ijsmart-1791-874X-24c.

Norberg, H. M., Maehle, N., & Korneliussen, T., (2011). From commodity to brand: Antecedents and outcomes of consumers' label perception. *Journal of Product and Brand Management, 20*(5), 368–378. https://doi.org/10.1108/10610421111157892.

Paul, D., (2003). *Developing the Brand: The Case of Alcohol, 1800-1880.*

Pennington, J., Ball, D., & Pennington, J. R., (2009). Customer branding of commodity products: The customer-developed brand power and effect size in marketing view project. *Article in Journal of Brand Management*. https://doi.org/10.1057/palgrave.bm.2550131.

Pitt, L. F., Watson, R. T., Berthon, P., Wynn, D., & Zinkhan, G., (2006). The penguin's window: Corporate brands from an open-source perspective. *Journal of the Academy of Marketing Science, 34*(2), 115–127. https://doi.org/10.1177/0092070305284972.

Punniyamoorthy, M., (2011). A framework for assessment of brand loyalty score for commodities. *Journal of Targeting, Measurement and Analysis for Marketing, 19*, 243–260. https://doi.org/10.1057/jt.2011.23.

Robinson, T., Clarke-hill, C. M., & Clarkson, R., (2002). *Differentiation Though Service: Service Industries Journal, 22*(3).

Romaniuk, J., & Sharp, B., (2004). *Conceptualizing and Measuring Brand Salience*. https://doi.org/10.1177/1470593104047643.

Shostack, G. L., (1989). Service blueprinting effectiveness: Drivers of success. *European Journal of Marketing, 16*(1), 580. https://doi.org/10.1108/EUM0000000004799.

Uncles, M. D., (2010). Market trends: Understanding retail customers. In: *Retailing in the 21st Century: Current and Future Trends* (2nd edn., pp. 205–219). Springer Berlin Heidelberg. https://doi.org/10.1007/978-3-540-72003-4_13.

Unger, L., (1983). Strategic planning for commodities and specialties. A strategic industry study based on the example of the chemical industry. *Long Range Planning, 16*(4), 12–20. https://doi.org/10.1016/0024-6301(83)90154-1.

Vertinsky, I., & Zhou, D., (1999). *The Economics of Certifying the Environmental Friendliness of Products*. https://doi.org/10.7939/R3J09W37Q (accessed on 8 December 2023).

Wengrow, D., (2008). Prehistories of commodity branding. *Current Anthropology, 49*(1), 7–21. https://doi.org/10.1086/523676.

CHAPTER 3

Influence of Personality Traits on the Purchase Behavior of Consumers in Kolkata

DEBARATI BANERJEE[1] and SUDIPTA MAJUMDAR[2]

[1]PhD Scholar, Faculty of Management Studies, ICFAI University, Jharkhand, India

[2]Associate Professor, Department of Management, School of Business & Economics, Adamas University, India

ABSTRACT

Shopping is a very important activity in everyone's life. Our life revolves around shopping for various utility materials as it is usually need-based. Sometimes shopping is also considered a leisure activity and acts as a form of entertainment for some people. Compulsive buying or purchasing behavior occurs when a condition leads to an excessive buying situation resulting in the uncontrollable acquisition of things despite significant psychological, social, vocational, and financial implications (CBB). When studying consumer buying behavior, it's important to remember that external stimuli aren't the only thing that causes a consumer to react; rather, the consumers' personality traits have a big impact on how they perceive stimuli, use their internal processes, and end up with a specific buying behavior. D. W. Fiske established the Big 5 Personality traits in 1949 which describe the theory of extraversion or extroversion, agreeableness, openness, conscientiousness, and neuroticism. The authors of this chapter investigated the impact of personality qualities on Kolkata consumers' purchase behavior during the COVID-19 pandemic in terms of the Big 5 Personality trait.

Sustainability in Marketing Practice: Strategies for Industry 4.0.
Subrata Chattopadhyay, Udit Chawla, (Eds.)

The majority of the responders are parents who have been persuaded by their children to engage in compulsive or impulsive purchasing behavior. The replies are analyzed to see which factors have a substantial impact on purchase behavior. The study's findings will aid the company in determining its marketing tactics.

3.1 INTRODUCTION

Consumers' buying behavior is a composite phenomenon that is driven by specific factors like situation, demography, characteristics, and personality traits of an individual. The purchasing patterns of consumers demonstrate how people obtain services from the market. However, the patterns are highly susceptible to change. Marketers are usually interested in the buying patterns and how they relate to specific market factors. Several comprehensive studies are carried out to better understand the buying trend and how it is related to the buyer's geographical, demographic, and psychological characteristics. This along with the Big 5 Personality traits could provide valuable insights. A better understanding of consumer buying behavior will enable companies to devise marketing strategies for market segmentation, distribution, and sales promotion.

An attribute of a person is defined as a distinctive and enduring characteristic that appears consistently across a range of situations and is not the same from one person to another. This indicates that an attribute of a person is quantitative and empirical. Certain personality traits, according to trait theorists, are extremely significant in understanding and studying consumer behavior. People could be graded on intelligence, assertiveness, emotional stability, and other factors (Table 3.1).

The following personality qualities are used to analyze consumer behavior specifically:

1. **Consumer Innovativeness (Parihar, 2017):** Consumers might be inventive or non-innovative depending on how open they are to new purchase experiences. Consumer innovativeness refers to people's desire to try new things. His dogmatism demonstrates his rigidity. Consumers with low dogmatism are more open-minded, whereas those with high dogmatism are more closed-minded.
2. **Consumer Susceptibility to Interpersonal Influence (Parihar, 2017):** Identifying the consumers who will respond well to social

influences (e.g., understanding how they will react to social influence). In addition to innovativeness, a second trait that is of importance is consumer susceptibility to interpersonal influence. It is possible to construct a scale to assess this trait.

3. **Consumer Materialism (Parihar, 2017):** It is to determine what is best for the consumer. Psychologists all over the globe agree that the Big Five Model, often known as the Five-Factor Model, is the most widely recognized theory of personality traits. Openness, conscientiousness, extraversion, agreeableness, and neuroticism, sometimes known as OCEAN or CANOE, are the five qualities (Diener et al., 2018).

TABLE 3.1 Description of Big Five Personality Traits

Big Five Traits	Definition
Openness	The inclination to favor sentiments and behaviors while appreciating new creative ideas.
Conscientiousness	The proclivity to be punctual for appointments, to obey rules, and to work diligently.
Extraversion	The proclivity to be chatty, friendly, and enjoy people; the proclivity to have a dominant style.
Agreeableness	The inclination to agree with others and follow their lead rather than expressing one's own thoughts and opinions.
Neuroticism	The proclivity for experiencing negative emotions such as wrath, concern, and grief on a regular basis, as well as being interpersonally sensitive.

There are two aspects of the five personality traits. Some people score low and some score high. Table 3.2 shows the different reactions to the 5 personality traits (Diener et al., 2018).

As we've seen, there are a variety of elements that influence a consumer's purchase pattern, both favorably and adversely. In this study, we'll look at personality qualities that influence a consumer's purchase pattern or buying behavior in Kolkata. The Big 5 Personality Traits variable was chosen after a thorough literature analysis. Each personality trait's perception by customers is being studied. This research intends to present a glimpse of consumers' purchase behavior in Kolkata during the COVID-19 pandemic in terms of various personality factors.

TABLE 3.2 Different Types of Reaction to the Personality Traits

Big Five Traits	Example Behavior for Low Scorers	Example Behavior for High Scorers
Openness	Prefers not to be exposed to various moral systems; has restricted interests; is inartistic; is not analytical; is practical.	Enjoys seeing people with unusual hairstyles and body piercings; inquisitive; unconventional; imaginative.
Conscientiousness	Unreliable; hedonistic; carefree; lax; prefers spur-of-the-moment action over planning.	Never late for a date; well-organized; diligent; neat; tenacious, punctual; self-disciplined.
Extraversion	Preferring a calm evening to a boisterous gathering; stoic.	Being the life of the party; being lively, cheerful, and fun-loving; and being affectionate.
Agreeableness	Asserts one's own rights quickly and confidently; angry; manipulative; obnoxious; unwilling to cooperate.	Pleasant natured; gullible; forgiving; helpful; agrees with others on political issues; good natured; gullible; forgiving; helpful.
Neuroticism	Not easily upset by little irritations; cool; unemotional; hardy; secure; self-assured.	Worrying excessively about minor details; uneasy; hypochondriacally; feeling inadequate.

3.2 LITERATURE REVIEW

Philip Thomas Udo-Imeh conducted a study on the impact of personality on undergraduate students' purchasing behavior in colleges in Cross River State. The 'big five' personality qualities - agreeableness, extraversion, openness to experience, conscientiousness, and neuroticism - were used to assess personality. Undergraduate students were the focus of the research. Personality is also a key predictor of customer purchasing behavior, according to the study. Agreeableness had the greatest influence on purchasing behavior among the five personality characteristics, while neuroticism had the least. The 'Big Five' personality attribute has the greatest level of agreement among scholars and researchers in the field of consumer behavior and is the most widely used personality assessment (Udo-Imeh, 2015).

Psychological segmentation as a marketing tactic can be employed when more traditional reasons for market division are exhausted, according to a study by Dr. N. Bindu Madhavi and Dr. P. Adi Lakshmi. Despite the fact that there is no data on the distribution of individual personality traits or composite personality profiles in a population, marketers can create product "personalities" to appeal to specific sorts of clients. If the target market's personality qualities are well identified, the company will be able to develop more successful marketing techniques to keep existing clients and convert them into satisfied customers (Madhavi & Lakshmi, 2020).

Ramgopal Ratnawat and Sachin Borgave researched the impact of Big-five personality factors on impulsive buying behavior on a random sample. While agreeableness and neuroticism were shown to be highly connected with impulsive buying behavior, extraversion, openness, and conscientiousness were found to have practically no relationships with impulsive buying behavior. These findings are in line with previous research (Borgave & Ratnawat, 2019).

In one of his studies, Prof. (Dr.) B. B. S Parihar found that personality had a significant impact on customer behavior. Consumer personality qualities such as consumer innovativeness, consumer susceptibility, and consumer materialism are substantially associated with demographic variables such as age, gender, income, and education of home appliance buyers in the majority of cases. In other words, the presence of the aforementioned consumer personality traits in a home appliance buyer is influenced by demographic factors such as age, gender, income, and education (Parihar, 2017).

Monika Fayez & Ashraf Labib in their research on big five personality on compulsive buying behavior they have found out that other than conscientiousness all other factors of big five scales have a positive impact on the compulsive buying behavior (Fayez & Labib, 2016).

Consumer impulsive purchase behavior is influenced by their gender and personality features. Khagendra Nath Gangai and Rachna Agrawal discovered this in one of their studies. The findings of the study will aid marketers in better understanding consumers' impulsive buying behavior and the personality qualities that influence their decision to make impulsive purchases (Nath & Agrawal, 2016).

According to one study, the buying intention in the case of augmented reality is significantly higher than in the case of traditional electronic commerce, indicating that a higher attractiveness of the system may lead not only to a more favorable attitude toward online shopping but also to a more intense intention to buy online. The findings backed up prior research on the subject that found links between personality factors and online shopping habits. The inclination to buy online was linked to extraversion and neuroticism (Lixăndroiu et al., 2021). The current research aims to identify and study the impact of the big 5 personality traits on consumers' buying patterns.

According to Christopher J Soto, a number of significant life consequences have been demonstrated. Agreeableness has been linked to a variety of affiliative and pro-social effects. Conscientiousness is a significant predictor of academic success and health outcomes. Subjective well-being and psychological health are adversely correlated with neuroticism. Extraversion reliably predicts social outcomes, while openness to experience is most strongly connected with intellectual and creative achievements (Soto, 2018).

The Five Factor Model is a variation of trait theory, a way of looking at the world that regards individual distinctions as the essence of human nature. Trait theory has been the dominant paradigm in European personality psychology since 1989 (Thomae, 1989), and a major component of American psychology since at least Allport's time (Buss, 1989). The assumptions of trait theory are inherent in most of psychometrics, and many researchers who would not identify themselves as trait theorists use them in this way. We anticipate finding consistent and enduring individual differences—the cardinal qualities of traits—which is why we normally seek internal consistency and retest reliability from scales (McCrae et al., 1992).

Deborah Cobb-Clark and Stefanie Schurer concluded in their study "The Stability of Big Five Personality Traits" that personality traits indeed appear to be stable among working-age people. Changes in the Big-Five personality traits at the mean level are minor and do not differ significantly between age groups. Furthermore, there is little evidence that a change in an individual's personality that is economically significant may be connected to negative work, health, or family events. Finally, these findings for the Big Five qualities are consistent with prior findings for locus of control, implying that non-cognitive talents may be viewed as reliable inputs in many economic decisions (Cobb-Clark & Schurer, 2011).

Anna Hermes and René Riedl's primary contributions to the paper "Influence of personality factors on retail purchase channel choice: Literature review and research agenda" are as follows: For starters, it expands our understanding of the effects of personality factors on channel selection behavior. The Big Five elements were found to be the most investigated in the context of online shopping in this study. According to some data, those who were high in agreeableness had fewer intents to shop online, whereas people who were high in openness were more likely to shop online. Furthermore, persons with a high level of neuroticism were shown to be more prone to shop online, although another study indicated the opposite. As a result, drawing broad conclusions about the direct effects of personality factors on channel selection should be approached with caution. The use of different measurement instruments, distinct demographic mark-ups, and diverse cultures may have resulted in inconclusive results (Hermes & Riedl, 2021).

3.3 RESEARCH METHODOLOGY

This is a research endeavor that is analytical in nature. It is concerned with the investigation of the impact of certain personality traits on consumer purchasing behavior. The consumers are mainly the Parents and Children displaying either Impulsive or Compulsive buying behavior which further are the dependent variables and the Big Five Personality Traits are the Independent Variables used in this study. For the collection of the data, we have approached several schools and 120 consumers in the Kolkata region, using a questionnaire as the data gathering tool. The analysis was made of 104 responses.

3.3.1 DEVELOPMENT OF QUESTIONNAIRE

In this study, we have used a structured questionnaire. Items for the structured questionnaire are gathered from the standardized scale of the Big Five personality. The items are referred to from mainly two literatures authored by Fayez & Labib (2016); and Ratnawat & Borgave (2019).

3.3.2 DATA COLLECTION METHOD

A questionnaire was utilized as an instrument, and a data collecting approach was used to get the needed information from primary sources. The responses for the independent variable were collected using the following five-point scale (McLeod, 2008): strongly disagree (SD) – 1; disagree (D) – 2; neither agree nor disagree (NAD) – 3; agree (A) – 4; and strongly agree (SA) – 5.

The responses for the dependent variable were collected using the following five-point scale (McLeod, 2008): not at all (NAA) – 1; sometimes (S) – 2; NAD – 3; very often (VO) – 4; and always (A) – 5.

3.3.3 ANALYSIS

One-way ANOVA is used to investigate the measurable effect on the dependent variables, and multiple regression is used wherein standardized regression coefficients (Beta) are used to measure how strongly each predictor variable influences the criterion variable. The higher the beta value, the greater the impact of the predictor variable on the criterion variable.

3.4 ANALYSIS

The psychographic variables which are studied are Big 5 Personality. One-way ANOVA is done in order to know whether personality traits have a significant influence on the consumers' (Parents and Children) buying behavior. The five predictor variables for big five personality traits have been identified, and data has been collected on them which are (Table 3.3):

1. **BFP1:** Consumers (Parents and Children) are very curious about new stuff launching in the market and always demand, the moment they are available in the market.
2. **BFP2:** Consumers (Parents and Children) buy stuff in a calculative manner after gathering feedback from the market.
3. **BFP3:** Consumers' (Parents' and Childrens') extravagant nature results in buying the wrong products.
4. **BFP4:** Consumers (Parents and Children) understand the value of money and therefore buy things selectively and spent money judiciously.
5. **BFP5:** Consumers (Parents and Children) sometimes suffer from depression which results in hasty shopping behavior.

The relevant piece of the SPSS result sheet is shown below to determine whether Big Five personality has a significant impact on customer purchasing behavior (Parents and Children).

TABLE 3.3 Index

Index in SPSS Output	Name of the Variable
BFP1 (Big five personality Dimensions 1 – Openness)	Big five personality dimensions [I am very curious about new stuff launching in the market and always demand, the moment they are available in the market] (SD – 1, D – 2, NAD – 3, A – 4, SA – 5).
BFP2 (Big five personality Traits 2 – Conscientiousness)	Big five personality traits [I buy stuff in a calculative manner after gathering feedback from the market] (SD – 1, D – 2, NAD – 3, A – 4, SA – 5).
BFP3 (Big five personality Traits 3 – Extraversion)	Big five personality traits [Extravagant nature of myself results into buying wrong products] (SD – 1, D – 2, NAD – 3, A – 4, SA – 5).
BFP4 (Big five personality Traits 4 – Agreeableness)	Big five personality dimension [I understand the value of money and therefore buy things selectively and spent money judiciously] (SD – 1, D – 2, NAD – 3, A – 4, SA – 5).

TABLE 3.3 *(Continued)*

Index in SPSS Output	Name of the Variable
BFP5 (Big five personality Traits 5 – Neuroticism)	Big five personality traits [Sometimes I suffer from depression which results in hasty shopping behavior (SD – 1, D – 2, NAD – 3, A – 4, SA – 5).
IP (Impulsive buying behavior of parents)	I am an impulsive buyer and buy products under impulsion? (Not at all – 1, Sometimes – 2, NAD – 3, Very often – 4, Always – 5).
IC (Impulsive buying behavior of children)	My child/children are the impulsive buyer and buy products under impulsion? (Not at all – 1, Sometimes – 2, NAD – 3, Very often – 4, Always – 5).
CP (Compulsive buying behavior of parents)	I am a compulsive buyer and buy products under compulsion? (Not at all – 1, Sometimes – 2, NAD – 3, Very often – 4, Always – 5).
CC (Compulsive buying behavior of children)	My child/children are compulsive buyers and buy products under compulsion? (Not at all – 1, Sometimes – 2, NAD – 3, Very often – 4, Always – 5).

3.4.1 BIG FIVE PERSONALITY TRAITS AND IMPULSIVE BUYING BEHAVIOR OF PARENTS

- **H_0:** Personality traits will not influence the Impulsive buying behavior of the parents.

  In the 6th Col. (Sig.) of Table 3.4, the exact significant level (p value) of ANOVA is displayed. We determined the level of significance to be 5%, or = 0.05 (on the basis of existing research of a similar type). Table 3.4 shows that the value of 'p' is less than the value of 'a.' The null hypothesis is rejected, and the alternative hypothesis is accepted, because p = 0.000 is less than 0.05. That is, personality factors have a considerable impact on parents' impulsive purchasing behavior.

TABLE 3.4 ANOVA Output for Big Five Personality Traits and Impulsive Buying Behavior of Parents

	Model	Sum of Squares	df	Mean Square	F	Sig.
1	Regression	19.855	5	3.971	5.024	0.000[a]
	Residual	74.305	94	0.790		
	Total	94.160	99			

[a] *Predictors*: (Constant), BFP5, BFP2, BFP1, BFP4, BFP3.

[b] *Dependent variable*: Impulsive Parents (IP).

Source: SPSS Output.

3.4.1.1 PRIORITIZATION OF THE VARIABLES WITH RESPECT TO THE SCALE OF BIG FIVE PERSONALITY TRAITS AND IMPULSIVE BUYING BEHAVIOR OF PARENTS

The Criterion Variable in this phase of the study is the parents' impulsive purchase behavior, for which five predictor factors related to Big Five Personality characteristics have been identified and data has been collected:

- **BFP1:** Consumers (impulsive parents) are very curious about new stuff launching in the market and always demand, the moment they are available in the market.
- **BFP2:** Consumers (impulsive parents) buy stuff in a calculative manner after gathering feedback from the market.
- **BFP3:** Consumers' (impulsive parents) extravagant nature results in buying the wrong products.
- **BFP4:** Consumers (impulsive parents) understand the value of money and therefore buy things selectively and spent money judiciously.
- **BFP5:** Consumers (impulsive parents) sometimes suffer from depression which results in hasty shopping behavior.

As previously stated, the goal of this section of the research is to identify the factor(s) that influence customers' behavior in the context of the big five personality traits. The relevant output from SPSS is shown in Table 3.5.

TABLE 3.5 Prioritization of the Variables with Respect to the Scale of Big Five Personality Traits and Impulsive Buying Behavior of Parents

	Model	Unstandardized Coefficients		Standardized Coefficients	*t*	Sig.
		B	Std. Error	Beta		
1	(Constant)	3.205	0.665		4.818	0.000
	BFP1	0.064	0.093	0.066	0.686	0.495
	BFP2	0.009	0.111	0.008	0.080	0.936
	BFP3	0.140	0.096	0.147	1.458	0.148
	BFP4	–0.487	0.130	–0.373	–3.732	0.000
	BFP5	0.063	0.106	0.061	0.593	0.554

[a] *Dependent variable Impulsive Parents (IP).*

Source: SPSS Output.

The following regression model is generated as a result of the output:

$$IP = 3.205 + 0.064BFP1 + 0.009BFP2 + 0.140BFP3 - 0.487BFP4 + 0.063BFP5$$

where; IP is the impulsive buying behavior of the parents.

The standardized regression coefficients (Beta) are a measure of how strongly each predictor variable affects the criterion variable, with the higher the beta value, the stronger the predictor variable's influence on the criterion variable.

BFP3 has the highest β value, 0.140, according to the model. It demonstrates that the aforementioned predictor variable has the greatest influence on the criterion variable. In reality, the variable, i.e., lavish nature or Extraversion, has a significant impact on Impulsive Parents, leading to the purchase of the incorrect products. Similarly, BFP4 has the lowest value, which is –0.487. It means the variable – 'Consumers' or Parents understand the value of money and therefore buy things selectively and spend money judiciously. Or Agreeableness' has the least impact.

3.4.2 BIG FIVE PERSONALITY TRAITS AND IMPULSIVE BUYING BEHAVIOR OF CHILDREN

- **H_0:** Personality traits will not influence the Impulsive buying behavior of the children.
  In the 6th Col. (Sig.) of Table 3.6, the exact significant level (p value) of ANOVA is displayed. We determined the level of significance to be 5%, or = 0.05 (on the basis of existing research of a similar type). Table 3.6 shows that the value of 'p' is less than the value of 'a.' The null hypothesis is rejected, and the alternative hypothesis is accepted, because p = 0.023 is smaller than 0.05. That is, personality factors have a major impact on children's impulsive purchasing behavior.

TABLE 3.6 ANOVA Output for Big Five Personality Traits and Impulsive Buying Behavior of Children

Model		Sum of Squares	df	Mean Square	F	Sig.
1	Regression	13.328	5	2.666	2.741	0.023[a]
	Residual	91.422	94	0.973		
	Total	104.750	99			

[a] *Predictors (Constant), BFP5, BFP2, BFP1, BFP4, BFP3.*

[b] *Dependent variable Impulsive Children (IC).*

Source: SPSS Output.

3.4.2.1 PRIORITIZATION OF THE VARIABLES WITH RESPECT TO THE SCALE OF BIG FIVE PERSONALITY TRAITS AND IMPULSIVE BUYING BEHAVIOR OF CHILDREN

The criterion variable in this phase of the study is children's impulsive purchase behavior, for which five predictor factors related to Big Five Personality traits have been identified and data has been collected:

1. **BFP1:** Consumers (impulsive children) are very curious about new stuff launching in the market and always demand, the moment they are available in the market.
2. **BFP2:** Consumers (impulsive children) buy stuff in a calculative manner after gathering feedback from the market.
3. **BFP3:** Consumers' (impulsive children) extravagant nature results in buying the wrong products.
4. **BFP4:** Consumers (impulsive children) understand the value of money and therefore buy things selectively and spent money judiciously.
5. **BFP5:** Consumers (impulsive children) sometimes suffer from depression which results in hasty shopping behavior.

As previously stated, the goal of this section of the research is to identify the factor(s) that influence customers' behavior in the context of the big five personality traits. The relevant output from SPSS is shown in Table 3.7.

TABLE 3.7 Prioritization of the Variables with Respect to the Scale of Big Five Personality Traits and Impulsive Buying Behavior of Children

	Model	Unstandardized Coefficients		Standardized Coefficients	t	Sig.
		B	Std. Error	Beta		
1	(Constant)	2.956	0.738	–	4.006	0.000
	BFP1	0.205	0.103	0.200	1.988	0.050
	BFP2	–0.006	0.123	–0.005	–0.045	0.964
	BFP3	0.095	0.107	0.095	0.894	0.373
	BFP4	–0.377	0.145	–0.273	–2.603	0.011
	BFP5	–0.136	0.118	–0.126	–1.154	0.251

[a] *Dependent variable Impulsive Children (IC).*

Source: SPSS Output.

The following regression model is generated as a result of the output.

$$IC = 2.956 + 0.205BFP1 - 0.006BFP2 + 0.095BFP3 - 0.377BFP4 - 0.136BFP5$$

where; IC is the Impulsive buying behavior of the children.

The standardized regression coefficients (Beta) are a measure of how strongly each predictor variable effects the criterion variable, with the higher the beta value, the stronger the predictor variable's influence on the criterion variable.

The highest value for BFP1 is 0.205, according to the model which is the β value. It demonstrates that the aforementioned predictor variable has the greatest influence on the criterion variable. In fact, the variable, i.e., 'open-mindedness' has a high level of impact on Impulsive Children resulting in buying new stuff launching in the market and always demanding them, the moment they are available in the market. Similarly, the β value for BFP4 is the lowest, i.e., –0.377. It means, the variable Consumers or Children understand the value of money and therefore buy things selectively and spend money judiciously, or Agreeableness has the least impact.

3.4.3 BIG FIVE PERSONALITY TRAITS AND COMPULSIVE BUYING BEHAVIOR OF PARENTS

- **H_0:** Personality traits will not influence the Compulsive buying behavior of the parents.

  In the 6th Col. (Sig.) of Table 3.8, the exact significant level (p value) of ANOVA is displayed. We determined the level of significance to be 5%, or = 0.05 (on the basis of existing research of a similar type). Table 3.8 shows that the value of 'p' is greater than the value of 'a.' The null hypothesis is accepted, while the alternative hypothesis is rejected, because p = 0.514 is greater than 0.05. That means personality traits does not impact the compulsive buying behavior of the parents.

TABLE 3.8 ANOVA Output for Big Five Personality Traits and Compulsive Buying Behavior of Parents

Model		Sum of Squares	df	Mean Square	F	Sig.
1	Regression	7.911	5	1.582	0.856	0.514[a]
	Residual	173.799	94	1.849		
	Total	181.710	99			

[a] *Predictors (Constant), BFP5, BFP2, BFP1, BFP4, BFP3.*

[b] *Dependent variable Compulsive Parent (CP).*

Source: SPSS Output.

3.4.3.1 PRIORITIZATION OF THE VARIABLES WITH RESPECT TO THE SCALE OF BIG FIVE PERSONALITY TRAITS AND COMPULSIVE BUYING BEHAVIOR OF PARENTS

The criterion variable in this section of the study is the Parents' Compulsive Buying Behavior, for which five predictor factors related to Big Five Personality Traits have been identified and data has been collected.

1. **BFP1:** Consumers (compulsive parents) are very curious about new stuff launching in the market and always demand, the moment they are available in the market.
2. **BFP2:** Consumers (compulsive parents) buy stuff in a calculative manner after gathering feedback from the market.
3. **BFP3:** Consumers' (compulsive parents) extravagant nature results in buying the wrong products.
4. **BFP4:** Consumers (compulsive parents) understand the value of money and therefore buy things selectively and spent money judiciously.
5. **BFP5:** Consumers (compulsive parents) sometimes suffer from depression which results in hasty shopping behavior.

As previously stated, the goal of this section of the research is to identify the factor(s) that influence customers' behavior in the context of the big five personality traits. The relevant output from SPSS is shown in Table 3.9.

TABLE 3.9 Prioritization of the Variables with Respect to the Scale of Big Five Personality Traits and Compulsive Buying Behavior of Parents

	Model	Unstandardized Coefficients		Standardized Coefficients	t	Sig.
		B	Std. Error	Beta		
1	(Constant)	2.885	1.017	–	2.836	0.006
	BFP1	–0.208	0.142	–0.155	–1.467	0.146
	BFP2	0.022	0.169	0.014	0.127	0.899
	BFP3	0.183	0.147	0.139	1.245	0.216
	BFP4	–0.115	0.200	–0.063	–0.575	0.566
	BFP5	–0.063	0.162	–0.044	–0.386	0.700

[a] *Dependent variable Compulsive Parent (CP).*

Source: SPSS Output.

The following regression model is generated as a result of the output.

$$CP = 2.885 - 0.208BFP1 + 0.022BFP2 + 0.183BFP3 - 0.115BFP4 - 0.063BFP5$$

where; CP is the compulsive buying behavior of parents.

The standardized regression coefficients (Beta) are a measure of how strongly each predictor variable effects the criterion variable, with the higher the beta value, the stronger the predictor variable's impact on the criterion variable.

The greatest value for BFP3 is 0.183, which is the β value according to the model. It demonstrates that the aforementioned predictor variable has the greatest influence on the criterion variable. In fact, the variable, i.e., 'extraversion' has a high level of impact on Compulsive buying behavior of Parents having extravagant nature resulting in buying the wrong products. Similarly, the β value for BFP1 is the lowest, i.e., –0.208. It means, the variable – or compulsive parents are very curious about new stuff launching in the market and always demand, the moment they are available in the market, or 'Openness' has the least impact.

3.4.4 BIG FIVE PERSONALITY TRAITS AND COMPULSIVE BUYING BEHAVIOR OF CHILDREN

- **H_0:** Personality traits will not influence the compulsive buying behavior of children.

  In the 6th Col. (Sig.) of Table 3.10, the exact significant level (p value) of ANOVA is displayed. We determined the level of significance to be 5%, or = 0.05 (on the basis of existing research of a similar type). Table 3.10 shows that the value of 'p' is greater than the value of 'a.' The null hypothesis is accepted, while the alternative hypothesis is rejected, because p = 0.365 is greater than 0.05. That means Personality traits does not impact the compulsive buying behavior of children.

TABLE 3.10 ANOVA Output for Big Five Personality Traits and Compulsive Buying Behavior of Children

Model		Sum of Squares	df	Mean Square	F	Sig.
1	Regression	8.712	5	1.742	1.101	0.365[a]
	Residual	148.728	94	1.582		
	Total	157.440	99			

[a] *Predictors (Constant), BFP5, BFP2, BFP1, BFP4, BFP3.*

[b] *Dependent variable Compulsive Children (CC).*

Source: SPSS Output.

3.4.4.1 PRIORITIZATION OF THE VARIABLES WITH RESPECT TO THE SCALE OF BIG FIVE PERSONALITY TRAITS AND COMPULSIVE BUYING BEHAVIOR OF CHILDREN

The criterion variable in this phase of the study is children's compulsive shopping behavior, which shows which five predictor factors related to Big Five Personality traits have been discovered and on which data has been collected:

1. **BFP1:** Consumers (compulsive children) are very curious about new stuff launching in the market and always demand, the moment they are available in the market.
2. **BFP2:** Consumers (compulsive children) buy stuff in a calculative manner after gathering feedback from the market.
3. **BFP3:** Consumers' (compulsive children) extravagant nature results in buying the wrong products.
4. **BFP4:** Consumers (compulsive children) understand the value of money and therefore buy things selectively and spent money judiciously.
5. **BFP5:** Consumers (compulsive children) sometimes suffer from depression which results in hasty shopping behavior.

As previously stated, the goal of this section of the research is to identify the factor(s) that influence customers' behavior in the context of the big five personality traits. The relevant output from SPSS is shown in Table 3.11.

TABLE 3.11 Prioritization of the Variables with Respect to the Scale of Big Five Personality Traits and Compulsive Buying Behavior of Children

Model		Unstandardized Coefficients		Standardized Coefficients	t	Sig.
		B	Std. Error	Beta		
1	(Constant)	2.410	0.941	–	2.560	0.012
	BFP1	–0.128	0.131	–0.102	–0.975	0.332
	BFP2	0.189	0.157	0.131	1.209	0.230
	BFP3	0.163	0.136	0.133	1.201	0.233
	BFP4	–0.245	0.185	–0.145	–1.326	0.188
	BFP5	0.013	0.150	0.010	0.089	0.929

[a] *Dependent variable Compulsive Children (CC).*

Source: SPSS Output.

The following regression model is generated as a result of the output:

$$CC = 2.410 - 0.128BFP1 + 0.189BFP2 + 0.163BFP3 - 0.245BFP4 + 0.013BFP5$$

where; CC is the compulsive buying behavior of children.

The standardized regression coefficients (Beta) are a measure of how strongly each predictor variable effects the criterion variable, with the higher the beta value, the stronger the predictor variable's influence on the criterion variable.

The greatest value for BFP2 is 0.189, which is the β value according to the model. It demonstrates that the aforementioned predictor variable has the greatest influence on the criterion variable. In fact, the variable, i.e., 'Conscientiousness' has a high level of impact on Compulsive buying behavior of Children buying stuff in a calculative manner after gathering feedback from the market. Similarly, the β value for BFP4 is the lowest, i.e., –0.245. It means, the variable – or Compulsive Children understand the value of money and therefore buy things selectively and spent money judiciously, or 'Agreeableness' has the least impact.

3.5 CONCLUSION

One-way ANOVA is used to determine whether the big five personality traits (Independent Variable) have a significant impact on the purchase pattern or buying behavior (impulsive or compulsive) of the parents and their children in order to meet the study's goal as stated in the introduction section of the chapter. The independent variables OCEAN which are considered in the study have a significant impact on the impulsive buying behavior or the purchase pattern of the parents and children. However, no such impact has been found on the compulsive buying behavior or on the purchase pattern of the parents and children.

In this study, multiple regression is also carried out to analyze the relationship between the single dependent variable and several independent variables. The following has been derived from the study:

- The variable, i.e., extravagant nature or Extraversion' has a high level of impact on Impulsive Parents resulting in buying the wrong products whereas Agreeableness' has the least impact.
- The variable, i.e., 'Open-mindedness' has high level of impact on Impulsive Children resulting in buying new stuff launching in the

market and always demanding them, the moment they are available in the market whereas Agreeableness' has the least impact.

- The variable, i.e., 'Extraversion' has a high level of impact on Compulsive Parents having extravagant nature resulting in buying the wrong products whereas Openness' has the least impact.
- The variable, i.e., 'Conscientiousness' has a high level of impact on Compulsive Children buying stuff in a calculative manner after gathering feedback from the market whereas Agreeableness' has the least impact.

In-depth research on facet-by-facet psychographic profiles on preference, on the other hand, may yield some extraordinary results that could be considered for future research.

KEYWORDS

- **analysis of variance**
- **big five personality traits**
- **compulsive buying behavior**
- **compulsive children**
- **COVID-19**
- **purchase pattern**
- **shopping behavior**

REFERENCES

Borgave, S., & Ratnawat, R., (2019). Impact of big-five personality on impulsive buying behavior article details abstract Article in International Journal of Multidisciplinary Research Review. https://www.researchgate.net/publication/335110080 (accessed on 8 December 2023).

Cobb-Clark, D., & Schurer, S., (2011). The Stability of Big-Five Personality Traits. Discussion Paper No. 5943 August 2011.

Diener, E., Lucas, R. E., & Cummings, J. A., (2018). Personality Traits Introduction to Psychology – Google Search. https://openpress.usask.ca/introductiontopsychology/chapter/personality-traits/#:~:text=This%20system%20includes%20five%20broad,grained%20analysis%20of%20someone's%20personality (accessed on 8 December 2023).

Fayez, M., & Labib, A., (2016). Investigating the effect of the "big five" personality dimensions on compulsive buying behavior of Egyptian consumers. In: A Journal of the Academy of Business and Retail Management (Vol. 10). ABRM. www.jbrmr.com (accessed on 8 December 2023).

Hermes, A., & Riedl, R., (2021). Influence of personality traits on choice of retail purchasing channel: Literature review and research agenda. In: Journal of Theoretical and Applied Electronic Commerce Research (Vol. 16, No. 7, pp. 3299–3320). MDPI. https://doi.org/10.3390/jtaer16070179.

Lixăndroiu, R., Cazan, A. M., & Maican, C. I., (2021). An analysis of the impact of personality traits towards augmented reality in online shopping. Symmetry, 13(3), 1–18. https://doi.org/10.3390/sym13030416.

Madhavi, N. B., & Lakshmi, P. A., (2020). Personality traits influencing the purchase decisions of passenger car buyers of Toyota in Krishna District. European Journal of Molecular & Clinical Medicine, 7(7).

Mccrae, R. R., John, O. P., Bond, M., Borkenau, P., Buss, D., Costa, P., Fiske, D., et al., (1992). An Introduction to the Five-Factor Model and Its Applications. 1992, pp. 175–215.

McLeod, S., (2008). Likert scale Likert scale examples how can you analyze data from a Likert scale? Simply Psychology, 1–3. https://www.simplypsychology.org/likert-scale.html (accessed on 8 December 2023).

Nath, G. K., & Agrawal, R., (2016). The Influence of Personality Traits on Consumer Impulsive Buying Behavior. http://www.publishingindia.com (accessed on 8 December 2023).

Parihar, B. B. S., (2017). Influence of selected personality traits on consumer behavior. In: Amity Journal of Management Amity Business School: (Vol. V, No. 1).

Soto, C., (2018). Big Five Personality Traits. https://www.researchgate.net/publication/324115204 (accessed on 8 December 2023).

Udo-Imeh, P. T., (2015). Influence of personality on the buying behavior of undergraduate students in universities in cross River State, Nigeria. International Journal of Marketing Studies, 7(4). https://doi.org/10.5539/ijms.v7n4p64.

PART II

Sustainable Business Strategy Practices

CHAPTER 4

A Conceptual Framework to Decode the Engagement and Satisfaction of Banking Consumers in Digital Space: A Review-Based Study

AJAY KUMAR GANGULY,[1] SUBRATA CHATTOPADHYAY,[1] and SUBHAJIT BHATTACHARYA[2]

[1]*Department of Management, University of Engineering and Management, Kolkata, West Bengal, India*

[2]*Department of Management, XIM University, Odisha, India*

ABSTRACT

This chapter provides information on the many reasons why people take actions in the banking sector, as well as some of the gaps that need to be filled in order to study this topic further. In this study, we use different models, techniques, and keywords to study how different aspects of technology use in the online banking sector are related. Researchers have found that the essential components of research on the adoption of online digital banking are scattered and incomplete. This means that it is difficult to understand how each factor affects happiness and loyalty among consumers who use digital banking. In this study, the authors looked at which concepts were used most frequently in literature from this particular period and also looked at which concepts were not being used as much in order to identify which areas might warrant further investigation. This study provides important guidance for those working in the banking sector on how to better understand the key elements of online banking services in order to better assess

Sustainability in Marketing Practice: Strategies for Industry 4.0.
Subrata Chattopadhyay, Udit Chawla, (Eds.)

customer satisfaction and loyalty. In order to improve our understanding of how consumers' behavior changes when they interact with banks, we have used a meta-analytical approach. This allows us to identify the research gaps and determine which factors are most predictive of consumer behavior.

4.1 INTRODUCTION

Information technology (IT) has evolved from a simple tool for automating preexisting operations to a critical decision-making tool and a critical component of providing services to clients (Diop, Zhao, & Duy, 2019). IT is now used worldwide. The banking, financial services, and insurance (BFSI) industry will feel the effects of this shift the most. IT has changed traditional business models and is becoming more important in the banking industry's efforts to improve customer service (Pradhan, Ghule, Yadav, & Shinde, 2020). By lowering entry barriers, enhancing competitiveness, and expediting the internationalization process, new technologies have a pivotal effect on banking capacity (Yadav, Ghule, & Shinde, 2020). Both internally and externally, technology has altered the banking industry in two ways: internally via the reorganization of banks' internal structure and production processes, and outside through product innovation and the introduction of new distribution networks (Lay, Boon, & Syaiful, 2016).

By the end of the 1980s, it was clear that the Indian banking industry needed computerization. Banks first used IT with the creation of stand-alone PCs, which were eventually upgraded to LAN connections. Banks eventually embraced the basic banking infrastructure as technology evolved. The transition from branch banking to bank banking was a watershed moment (Jayachandran, 2019). Core banking solutions allowed banks to improve consumer comfort, and it was seen as a promising step toward "Anywhere and Anytime Banking," which promised to enhance customer convenience (Rajan & Shamini, 2018). In the early 1990s, when the economy opened up, the digitization process picked up speed. This shift was sped further by rising competition among local and foreign banks. More than a few commercial banks have started offering digital consumer services to stay competitive.

In 2016, the government established the unified payments interface (UPI), a banking system that allows customers to move money across bank accounts with ease (Rastogi, Panse, Sharma, & Bhimavarapu, 2021). As a result of the UPI system, India saw a digital revolution and the advent of "Mobile Banking" (Kumar, Dhingra, Batra, & Purohit, 2020). Since it was not based on mobile technology, this kind of digital money transfer wasn't accessible by all account holders before 2016. Anyone may now use a mobile phone to register for a bank account and make transactions.

Banks started researching methods to link their multiple sites to communicate information as early as the 1990s (De Oliveira & Hwang, 2018). Private banks and international organizations with varied goals and large IT investments have boosted e-banking in India (Shahabi & Faezy, 2019). Innovators in financial technology were able to supply a new range of financial products and services at significantly lower costs. At the early stage, the network and backend infrastructure (which includes servers, workstations, networks, and other components) were underdeveloped. Infrastructure (data centers, bandwidth, etc.) was distributed widely. Another way of saying this is that every branch has its own rules. Servers, banking applications, databases, and several other sorts of systems. A variety of people and support components were affected by this structure, which resulted in problems with updating and debugging, as well as operational and budgetary challenges (Aduda & Kingoo, 2012). With the advent of new distribution channels like the internet and mobile, it became necessary to access and update the information in real-time from any branch or channel. As rapidly as possible, the networks had to be up and running, and they also had to be handled effectively. Authors have realized the benefits and drawbacks of scalability and scope; therefore, the architecture of data centers has become more centralized (Alizadeh et al., 2010). As a result, users, the network's administration, and network operators have had a lot of help in modernizing and simplifying the network. The introduction of core banking in India has resulted from the centralization of technological infrastructure in the banking domain (Krishnamoorthy & Shivkumar, 2020).

Following Oliver's (1981) research, which offered a design for a concept fundamental to the understanding of consumer behavior, the notion of customer satisfaction became widespread in the marketing sector. McCullough et al. (1986) performed a groundbreaking study on customer satisfaction in the banking industry. Satisfaction is a mental state that occurs when performance results and expectations are compared in the context of a customer's experience (Oliver, 1981). As a cognitive or an emotive process, satisfaction might be defined as a feeling of contentment (Mano & Oliver, 1993). The satisfaction concept has been examined in numerous ways over the last several decades based on these definitions. Many theories, approaches, and paradigms have consolidated the aspects that lead to contentment.

Three factors have contributed to the increased importance of customer satisfaction investigations in the banking sector (DeLone & McLean, 1992). For starters, there has been a plethora of research conducted on the measuring of client satisfaction with financial institutions. Second, there are legitimate ways for determining this construct's presence or absence. Apart from that, customer happiness is crucial to the achievement of long-term objectives and is specifically linked with the success of businesses (Cooil, Keiningham, Aksoy, & Hsu, 2007). Studies on customer satisfaction with banks have been conducted in a variety of situations, services, and nations throughout the years. Several research has looked at the factors that influence and are influenced by contentment. Studies have looked at the factors that influence consumer satisfaction in online environments, including ease of use (Li, Lu, Hou, Cui, & Darbandi, 2021), convenience (Ibrahim, Joseph, & Ibeh, 2006), accessibility (Babakus & Boller, 1992), privacy (Ladhari, 2009), and security (Asubonteng, McCleary, & Swan, 1996), reliability (Teeroovengadum, 2020), responsiveness (Iberahim, Taufik, Adzmir, & Saharuddin, 2016). Furthermore, constructs such as perceived quality (Mansour, Kooli, & Utama, 2014); perceived value (Durkin, O'Donnell, Mullholland, & Crowe, 2007) and satisfaction (Kaura, 2013) with the issue of consumer perspective have been studied in this chapter. The consumer expectation fulfillment outcomes that have been studied with respect to trust (Alalwan, Dwivedi, & Rana, 2017), consumer satisfaction (Li, Lu, Hou, Cui, & Darbandi, 2021), behavioral intention and attitude (Guriting & Ndubisi, 2006) and loyalty (Bowen & Chen, 2001).

Within the scope of this research, an attempt was made to identify the most significant constructs that are both directly and indirectly associated with consumer behavior in the online digital banking service sector. In addition, the goal of this study is to identify which combination(s) of the various primary factors that characterize consumers' opinions and expectations on online and digital banking, as reported in the present literature, may be the most effective combination of those factors. This research is based on a thorough evaluation of the current literature in the field of online and digital banking services, which was conducted for this purpose to determine the most appropriate technique for explaining consumer behavior connected to online and digital banking. Another focus of this research was grouping key variables linked with online digital banking to understand consumer behavioral intentions. In addition to it, this research also tried to focus on qualitative and quantitative aspects of different independent and dependent variables and their interrelations.

This chapter presents a systematic literature review (SLR) approach to distinguish between distinct predictor determinants of customer behavioral outcomes in the banking sector and to unearth the research gap, which may lead to identifying the scope of future research. Using a systematic meta-analytical review, this study has attempted to reveal the various dimensions of different articles on online banking, most of which have been published in Scopus-indexed journals, as well as the potential future research areas that could serve as a springboard for future work. This review-based chapter attempted to find out the key variables and the appropriate methodological approach to interpret the different outcomes related to consumer behavior in the domain of online and digital bank marketing by conducting a quantitative and qualitative measurement of all the published papers.

4.2 THEORETICAL FRAMEWORK

A literature review is critical for distilling the substance of publications, establishing facts, assessing the content, cataloging, and identifying other publications on the topic (Shergill & Li, 2005). Another advantage of the systematic review analysis is that it may assist in identifying research gaps in the new topic and provide recommendations for further investigations (Tong, Wong, & Lui, 2012). The previous structured approach cannot handle the huge volumes of data required to provide a scientific foundation for a research study, while thorough comprehensive literature analysis is capable of doing it (Zhu, Yan, Yang, & Xu, 2021). Observations have been made in the field of bank marketing that internet banking, digital banking, and mobile banking are becoming more popular as areas of academic interest (Kaura, 2013).

4.2.1 ANTECEDENTS OF CONSUMER SATISFACTION IN ONLINE BANKING

Although the number of publications continues to grow, the academic literature continues to present issues that need to be addressed, as seen by the large number of results that are in dispute with one another. Taking the example of dependability or reliability and pleasure or satisfaction, the relationship between these two constructs might be negative (Unyathanakorn, Kasikornbank, & Rompho, 2014) positive (Yoon, 2010), or even neutral in some literature (Li, Lu, Hou, Cui, & Darbandi, 2021) depending on the outcomes and circumstances. Other correlations, looking at the other antecedents, also show these variations. As a consequence of the discrepancies in the findings, it is recommended of doing a SLR study to synthesize and generalize the key findings on online banking service satisfaction, and then prepare a systematic review that summarizes research on the constructs and consequents of customer satisfaction and loyalty in the digital banking domain (Arbore & Busacca, 2009). Six of the 10 primary digital banking trends are presently tied to the digital environment via the rapid deployment of digital distribution, smartphone design, more digital and social selling, broad market adoption of mobile payments, and the growing effect of digital insurgents.

As part of their study of 19 service quality models, Seth et al. (2005) assessed their findings critically (1984–2003); studies on the development of e-service quality measuring scales published between 2001 and 2008 were evaluated by Ladhari (2010) using a systematic analysis of research that looked at the relationship between trust and online banking adoption, Goudarzi et al. (2013) revealed that trust had a positive relationship with adoption of internet banking. Trust is a critical aspect in online financial transactions, according to the authors' (Muñoz-Leiva, Climent-Climent, & Liébana-Cabanillas, 2017). Numerical trust, interpersonal trust, and organizational trust were recognized by McKnight & Chervany (2001) as the three categories of trust. There are three phases of trust formation, according to the researchers: building, consistency, and disintegration.

The evaluations listed above were primarily concerned with three characteristics of the online banking domain: quality dimensions, diffusion, and adoption, and reliability and trust. However, there is a dearth of detailed literature reviews on the subject of online banking. Customer satisfaction demands a multidisciplinary approach in online banking that integrates concepts from marketing, finance, consumer behavior, information, and communication technology, performance measurement, and management. Our study seeks to bridge that gap by examining research on online digital banking that was published between 1997 and 2021 in Scopus indexed journals from a variety of disciplines. The different extracted antecedents and their short descriptions have depicted Table 4.1.

TABLE 4.1 Different Extracted Antecedents and Their Short Descriptions

Antecedents	Short Description	References
Ease of use	User-friendliness in obtaining information and services in the digital context.	Li et al. (2021); Mehrad & Mohammadi (2017); Rodrigues, Oliveira, & Costa (2016); Sinha & Mukherjee (2016); Muñoz-Leiva, Climent-Climent, & Liébana-Cabanillas (2017); Aldás-Manzano, Lassala-Navarré, Ruiz-Mafé, & Sanz-Blas (2009); Patel & Patel (2018)
Reliability	Accuracy and dependability in providing the service that the consumer expects.	Zhou (2004); Teeroovengadum (2020); Yilmaz, Ari, & Gürbüz (2018); Shankar & Jebarajakirthy (2019); Dauda & Lee (2016); Kant & Jaiswal (2017); Chakrabarti, Trehan, & Makhija (2018); Narteh (2018); Ayo, Oni, Adewoye, & Eweoya (2016); Hamzah, Lee, & Moghavvemi (2017)
Content of information	Accuracy of material aimed towards users' interests and demands.	Pikkarainen et al. (2006); Yoon (2010); Ahmad & Al-Zı′bi (2011); Zhu & Chen (2012); Marakarkandy & Yajnik (2013)
Convenience	Content is easily accessible and robust in respect of consumer.	Joseph, Sekhon, Stone, & Tinson (2005); Joseph, McClure, & Joseph (1999); Kaura, Prasad, & Sharma (2015); Dauda & Lee (2016); Wasan (2018)
Website design	Attractiveness and pleasure of the virtual environment	Shankar & Jebarajakirthy (2019); Bashir & Madhavaiah (2015); Yoon (2010); Ramseook-Munhurrun & Naidoo (2011); Ahmad & Al-Zı′bi (2011); Zhu & Chen (2012); Momeni et al. (2013)
Efficiency	Overall performance of the system and accessibility to banking services.	Joseph, Sekhon, Stone, & Tinson (2005); Bloemer, de Ruyter, Ko, & Peeters (1998); Joseph, McClure, & Joseph (1999); Amin (2016); Raza, Umer, Qureshi, & Dahri (2020); Hammoud, Bizri, & El-Baba (2018); Allada & Dubey (2014)
Assurance	Accuracy in terms of delivering the service as promised.	Santouridis & Kyritsi (2014); Roche (2014); Joseph, Sekhon, Stone, & Tinson (2005); Yilmaz, Ari, & Gürbüz (2018); Dauda & Lee (2016); Hossain, Dwivedi, & Naseem (2015); Tetteh (2020)
Trust on bank image	A strong conviction in someone or something's dependability, veracity, or ability.	Mukherjee & Nath (2003); Nguyen (2020); Sharma, Govindaluri, Muharrami, & Tarhini (2017); Mulia, Usman, & Parwanto (2020); Hamzah, Lee, & Moghavvemi (2017); van Esterik-Plasmeijer & van Raaij (2017); Arcand, PromTep, Brun, & Rajaobelina (2017); Marinkovic & Obradovic (2015); Mehrad & Mohammadi (2017)
Service quality	Refers to a customer's evaluation of service expectations in relation to a company's performance.	Chai, Tan, & Goh (2016); Yavas, Bilgin, & Shemwell (1997); Zameer et al. (2015); Özkan et al. (2019); Levy & Hino (2016); Marinkovic & Obradovic (2015); Amin (2016); Makanyeza & Chikazhe (2017); Saleem et al. (2016); Mansour et al. (2016); Shayestehfar & Yazdani (2019); Islam et al. (2020); Eklof, Podkorytova, & Malova (2020)

TABLE 4.1 *(Continued)*

Antecedents	Short Description	References
Responsiveness	The ability to respond positively and promptly to a situation.	Chi Cui, Lewis, & Park (2003); Yavas, Bilgin, & Shemwell (1997); Zhou (2004); Yilmaz, Ari, & Gürbüz (2018); Dauda & Lee (2016); Kant & Jaiswal (2017); Chakrabarti, Trehan, & Makhija (2018); Narteh (2018); Ayo, Oni, Adewoye, & Eweoya (2016); Kashif, AbdurRehman, & Pileliene (2016); Tetteh & Boachie (2020)
Empathy	Having the capacity to recognize and experience the emotions of another.	Tetteh & Boachie (2020); Ahmed, Vveinhardt, Štreimikienė, Ashraf, & Channar (2017); Almugari, Bajaj, Tabash, Khan, & Ali (2020); Hossain, Dwivedi, & Naseem (2015); Ali & Raza (2017); Torres Fragoso, & Luna Espinoza (2017)
Security	The condition of not being in danger or under peril.	Hammoud, Bizri, & El-Baba (2018); AlNajem (2018); Almugari, Bajaj, Tabash, Khan, & Ali (2020); Yu (2008); Gupta, Yun, Xu, & Kim (2017); Torres Fragoso, & Luna Espinoza (2017); Jibril, Kwarteng, Botachway, Bode, & Chovancova (2020)
Risk	A situation in which one is exposed to risk.	Santini, de Ladeira, Sampaio, Perin, & Dolci (2020); Owusu, Bekoe, Addo-Yobo, & Otieku (2020); Gupta, Yun, Xu, & Kim (2017); Nguyen (2020); Teka (2020); Krishna & Sequeira (2016); Bashir & Madhavaiah (2015); Hanafizadeh & ZareRavasan (2017); Alalwan, Dwivedi, Rana, & Williams (2016); Giovanis, Athanasopoulou, Assimakopoulos, & Sarmaniotis (2019)
Tangibility	Capable of being sensed, particularly via the sense of touch.	Hinson, Owusu-Frimpong, & Dasah (2011); Chi Cui, Lewis, & Park (2003); Kant & Jaiswal (2017); Chakrabarti, Trehan, & Makhija (2018); Narteh (2018); Hamzah, Lee, & Moghavvemi (2017); Kashif, AbdurRehman, & Pileliene (2016); Tetteh & Boachie (2020)
Demographic variables	Demographics is a term that relates to the description or distribution of a target audience, client base, or population's characteristics.	Al-Jazzazi, & Sultan (2017); Choudrie, Junior, McKenna, & Richter (2018); Sharma, Govindaluri, Muharrami, & Tarhini (2017); Allada & Dubey (2014)
Customer service and support	Customer support is a collection of services that provide to assist customers in getting the most and resolving the issues.	Hinson, Owusu-Frimpong, & Dasah (2011); Joseph & Stone (2003); Shankar & Jebarajakirthy (2019); Paul, Mittal, & Srivastav (2016); Joseph, McClure, & Joseph (1999); Wasan (2018)
Social influence	Social influence involves both deliberate and accidental attempts to alter another person's views, attitudes, or conduct.	Akhtar, Irfan, Sarwar, Asma, & Rashid (2019); Almugari, Bajaj, Tabash, Khan, & Ali (2020); Santini, de Ladeira, Sampaio, Perin, & Dolci (2020)

Source: Authors.

4.2.2 CONSEQUENTS OF CONSUMER SATISFACTION IN ONLINE BANKING

The authors look at the behavioral intention (Alalwan, Dwivedi, & Rana, 2017), consumer satisfaction (Tong, Wong, & Lui, 2012), and loyalty (Mahamad & Ramayah, 2010) constructs in order to determine consumers' attachment. It is believed that there would be a positive association between customer satisfaction and loyalty since satisfied customers will be more likely to repurchase and spread good word of mouth, as confirmed by research relevant to this topic (Thakur, 2014). The different consequents in layer format extracted from different articles under the purview and their short descriptions have depicted Table 4.2.

TABLE 4.2 Different Extracted Consequents and Their Short Descriptions

Consequents	Short Description	References
Behavioral intention [Layer 1]	Refers to the motivating elements that impact a certain action, where the greater the desire to conduct the activity, the more likely it will be accomplished.	Saji & Paul (2018); Krishna & Sequeira (2016); Akhtar & Rashid (2019); Santini, Ladeira, Sampaio, Perin, & Dolci (2020); Jibril, Kwarteng, Botachway, Bode & Chovancova (2020)
Consumer satisfaction [Layer 2]	Overall happiness which conforms positive feelings.	Hamzah, Lee, & Moghavvemi (2017); Raza, Umer, & Dahri (2020); Tetteh & Boachie (2020); Islam & Al-Asheq (2020); Ahmed, Vveinhardt & Channar (2017); Hammoud & El Baba (2018); Allada & Dubey (2014); Eklof & Malova (2020)
Loyalty [Layer 3]	A continuing emotional bond.	Tetteh & Boachie (2020); Činjarević & Avdić (2010); Tetteh (2020); Eklof & Malova (2020); Kaura (2013)

Source: Authors.

4.3 METHODOLOGIES

4.3.1 DESIGN OF THE CONTEXTUAL REVIEW

Each review synthesizes prior research and assesses the strategies used to identify, select, and analyze key research as well as extract and direct future research. SLR articles may be classified as domain-based, theory-based, or method-based (Prabowo, 2016). Domain-based reviews are structured evaluations that concentrate on commonly used approaches, key concepts, and constructs (Alkhowaiter, 2020). Analyze and demonstrate how the frameworks operate together to read literature in 360°. Other review types include hybrid, meta-analysis, and bibliometric (Wu, 2017). This research is a zone-based hybrid structured review that tries to analyze techniques, variables extraction, keyword extraction, major models used in previous studies and key methodologies extraction which helps to identify the different factors in Technological adoption in Banking.

This SLR has five stages (Thorpe, Holt, Macpherson, & Pittaway, 2005). It was initially decided on the selection keys based on the paper's main theme. Step 2 look for articles in the database. Step 3 comprise extracting the abstracts from a number of sources. Step 4 leads to the review and choosing whole papers from a pool. And the final step depicts analyzing the research for zones, techniques, tools, and models as well as factors. The Scopus database discovered journals with the keywords mobile banking and internet online banking.

4.3.2 REVIEW STRUCTURE

4.3.2.1 SEARCHING FOR THE APPROPRIATE MATERIAL

This study's systematic review required two search techniques: A. *Electronic Search*: A list of relevant keywords was compiled based on the authors' experience and current online banking articles. In this case, a text mining approach was adopted. B. *Manual Search*: To booster our search method, the authors manually searched the articles' reference lists (i.e., the second search strategy employed in this research). This approach was used to check for any omissions throughout the list-making process (Rhaiem & Amara, 2019).

4.3.3 ARTICLES INCORPORATION CRITERIA

To limit the scope of our research technique and follow-up investigation, the authors devised include-and-exclude criteria which include the Indexing criteria, Redundant Criteria. The third and more important one is accessibility and availability criteria. After addressing all the above criteria, the authors include 158 articles. Here worth to be mentioned that the authors only considered Scopus Indexed journal articles.

4.3.4 COMPONENTS INCLUSION

Following earlier research studies on a systematic review of the literature (Ismagilova, Slade, Rana, & Dwivedi, 2019), keyword search was found adequate to meet this research's goal. The current study employed the main keywords to search Scopus for relevant research outputs. In the present study, the authors tried to include the key components for analyzing their usage through text mining software from the previous articles and tried to find out the domain (by identifying the keywords) where there is a chance to explore further research. In this respect, the antecedents and consequent constructs were also explored which helped them to identify the future scope of research. In this chapter, the authors have tried to study the models and methodologies which the previous researchers used and tried to identify the unexplored area for future research.

4.4 RESULT AND DISCUSSION

4.4.1 *TIME ZONE-BASED STUDIES INCLUDED*

Before evaluating papers, it is critical to understand the year wise breakdown of studies (Ahmad, Naveed, Ahmad, & Butt, 2020). Table 4.3 indicates that we took these articles from 1997 to 2021; the majority of the articles referenced are from 2017 to 2021 (around 101 articles out of 158 articles). Table 4.3 depicts the year-wise breakdown of the procured articles.

TABLE 4.3 Year-Wise Publication

Year	Number of Publication
2017–2021	101
2010–2015	41
2005–2009	5
2000–2004	8
1997–1999	3

Source: Authors.

4.4.2 *THEORIES AND MODELS USED*

Digital banking research has used many theories and models to identify users' intentions and use. Research on digital online Banking most often utilized UTAUT. Other factors impacting adoption behavior were added to this idea in most research.

TAM was another popular model. Some investigations combined two or more models. For example, Alshare & Mousa (2014) provided a cornerstone that studied the influence of proclaimed cultural values on customer intention to utilize online digital banking services. Research on digital banking utilized

TAM the most often. TAM was developed to examine digital banking adoption and usage. Studies using SERVQUAL and e-SERVQUAL to examine customer satisfaction aspects have been done in the past as well. Reliability, tangibility, responsiveness, and security are the most often utilized dimensions of SERVQUAL. The spread of innovation models was employed in another set of investigations.

In other cases, the research framework was built on the theoretical foundations of more than one theory and model. There are several examples where the e-SERVQUAL and IPA framework has combinedly used. The same thing has been seen in SERVCON and RATER models. Table 4.4 depicts a tabular thematic representation of different models obtained under this review work.

TABLE 4.4 Different Models Obtained from the Articles Under the Purview of This Systematic Literature Review [SLR] Work

Model Used	References
UTAUT2	Alalwan, Dwivedi, & Rana (2017)
ICA & NSBM	Lin & Chiu (2013)
TAM & Extended TAM	Sinha & Mukherjee (2016); Muñoz-Leiva, Climent-Climent & Liébana-Cabanillas (2017); Aldás-Manzano Lassala-Navarré, Ruiz-Mafé, & Sanz-Blas (2009); Patel & Patel (2018); Ben (2016); Alalwan & Williams (2016); Bashir & Madhavaiah (2015); Mulia & Parwanto (2020); Sharma & Tarhini (2017); Saji & Paul (2018); Teka (2020); Owusu & Otieku (2020); Santini & Dolci (2020); Ahmad & Hwang (2020)
DOI	Sinha & Mukherjee (2016)
SERVCON Framework	Rodrigues & Costa (2016); Ben (2016); Alalwan & Williams (2016); Sharma & Tarhini (2017); Shankar & Rishi (2020); Kaura (2013); Kaura & Sharma (2015)
SERVQUAL/e-SERVQUAL Framework	Hinson & Dasah (2011); Chi & Park (2003); Yilmaz & Gürbüz (2018); Chakrabarti & Makhija (2018); Narteh (2018); Hamzah & Moghavvemi (2017); Raza & Dahri (2020); Tetteh & Boachie (2020); Ahmed, Ashraf, & Channar (2017); Choudhury (2014); Ali & Raza (2017); Yu (2008); Činjarević, Tatić, & Avdić (2010); Allada & Dubey (2014); Ahmad & Hwang (2020); Hammoud & El-Baba (2018); Shayestehfar & Yazdani (2019); Ayo & Eweoya (2016); Amin (2016); Kant & Jaiswal (2017); Shankar & Jebarajakirthy (2019); Yavas & Shemwell (1997)
SERVPERF or RATER Framework	Torres & Luna (2017); Chi & Park (2003); Zhou (2004); Chakrabarti & Makhija (2018)
IPA Model	Joseph & Stone (2003); Ringle & Sarstedt (2016); Yu (2008)
Onion Research Model	Al-Najem (2018)

Source: Authors.

4.4.3 *METHODOLOGIES USED*

Previous research employed a variety of qualitative and quantitative methodologies to obtain data about digital banking and payment systems. Structured equation modeling (SEM), partial least squares (PLS), exploratory factor analysis and correlation and stepwise regression are the key analytical tools. Regression methods were used in the majority of online banking research studies. In the last two decades, structural equation modeling (SEM) has been the approach of choice for many academics looking into the intricate relationships between latent components (Astrachan, Patel, & Wanzenried, 2014). Of late, covariance-based SEM (CB-SEM) has been more popular and successful due to the rising difficulties of the approach in terms of distribution assumptions, sample size, and model complexity (Astrachan, Patel, & Wanzenried, 2014). It is most common to use a cross-sectional design when working with data from a diverse population. Longitudinal and panel designs are nonexistent in marketing and banking, suggesting that these strategies cannot be used. Table 4.5 depicts a glimpse of different methodologies used in the reviewed articles with proper references.

TABLE 4.5 Different Methodologies Used in Different Articles Under the Purview of the SLR

Key Methodologies Adopted	References
Structural equation modeling [SEM] [Path analysis/latent variable structural model/growth curve model].	Li, Lu, Hou, Cui, & Darbandi (2021); Teeroovengadum (2020); Muñoz-Leiva, Climent-Climent, & Liébana-Cabanillas (2017); Yilmaz, Ari, & Gürbüz (2018); Shankar & Jebarajakirthy (2019); Kant & Jaiswal (2017); Levy & Hino (2016); Khan, Rahman, & Fatma (2016); Narteh (2018); Patel & Patel (2018); Saleem, Zahra, Ahmad, & Ismail (2016); Ringle & Sarstedt (2016); Saji & Paul (2018); Hammoud, Bizri, & El-Baba (2018); Almugari, Bajaj, Tabash, Khan, & Ali (2020); Tetteh (2020); Teka (2020); Owusu, Bekoe, Addo-Yobo, & Otieku (2020); Kaura (2013); Mehrad & Mohammadi (2017); Zhou (2004); Özkan, Süer, Keser, & Kocakoç (2019)
Extended unified theory of acceptance and use (UTAUT2)	Alalwan, Dwivedi, & Rana (2017)
Dynamic panel data methods.	Akhisar, Tunay, & Tunay (2015)
ICA and NSBM	Lin & Chiu (2013)
Random utility model and multinomial logit model and rank-ordered logit model.	Yusuf & Lee (2015)

TABLE 4.5 *(Continued)*

Key Methodologies Adopted	References
Cross-validation approach and correlation, stepwise regression procedure [multiple regression/ multivariate regression/hierarchical regression].	Arbore & Busacca (2009); Sinha & Mukherjee (2016); Bloemer, de Ruyter, & Peeters (1998); Zameer, Tara, Kausar, & Mohsin (2015); Kaura, Prasad, & Sharma (2015); Mahmoud, Blankson, Owusu-Frimpong, Nwankwo, & Trang (2016); Mulia, Usman, & Parwanto (2020); Anouze & Alamro (2019); Sharma, Govindaluri, Muharrami, & Tarhini (2017); Akhtar, Irfan, Sarwar, Asma, & Rashid (2019); Azad (2016); Choudhury (2014); Mukherjee, Nath, & Pal (2002); Paul, Mittal, & Srivastav (2016)
Logistic regression model	Chakrabarti, Trehan, & Makhija (2018); Islam, Ahmed, Rahman, & Al Asheq (2020).
ANOVA	Hinson, Owusu-Frimpong, & Dasah (2011); Joseph & Stone (2003); Sayani (2015)
Exploratory/confirmatory factor analyses	Chi Cui, Lewis, & Park (2003); Hway-Boon & Ming Yu (2003); Kaura, Prasad, & Sharma (2015); Levy & Hino (2016); Wasan (2018); Khan, Rahman, & Fatma (2016); Patel & Patel (2018); Marinkovic & Obradovic (2015)
Conjoint analysis	Dauda & Lee (2016)
Artificial neural network	Anouze & Alamro (2019); Sharma, Govindaluri, Muharrami, & Tarhini (2017)
Partial least squares (PLS) technique	Hanafizadeh & ZareRavasan (2017); Raza, Umer, Qureshi, & Dahri (2020); Tetteh & Boachie (2020); Adapa & Roy (2017); Hossain, Dwivedi, & Naseem (2015)

Source: Authors.

4.4.4 ANTECEDENTS AND CONSEQUENTS IN ONLINE BANKING

It was revealed that numerous effects have been used in this literature analysis to analyze the implications of online banking, despite this section's emphasis on antecedents and consequents. Satisfaction, attitude, intention, and behavior patterns are all often attributed to a range of elements indifferent outcomes.

4.4.4.1 FACTORS IMPACTING BEHAVIORAL INTENTION IN DIGITAL BANKING

Several research indicated that constructs including trust (Mehrad & Mohammadi, 2017), perceived security (Damghanian, Zarei, & SiahsaraniKojuri, 2016), and usefulness (Hinson, 2011) mostly influence adoption behavior. Also, only a few research examined the impact of age, gender, education, and income (Dinçer, Yüksel, Eti, & Tula, 2019) on adoption intention. These demographic constructs had very limited influence on the intention to use digital banking services. A limited number of research discovered substantial correlations between online banking behavioral intention and effort anticipation (Demirgüç-Kunt, & Detragiache, 2000), self-efficacy (Rahayu, Rasid, & Tannady, 2018), security (Kurmanova, Ajupov, Abelguzin, & Kurmanova, 2016) and enabling circumstances.

4.4.4.2 FACTORS INFLUENCING DIGITAL BANKING ADOPTION

Very little numbers of research looked at the aspects that influence customers' actual user behavior. It was discovered that effort expectation (Allada & Dubey, 2014), social influence (Akhtar, Irfan, Sarwar, Asma, & Rashid, 2019) had little effect on online digital banking usage.

4.4.4.3 FACTORS IMPACTING EXPECTATION FULFILMENT AND SATISFACTION ON ONLINE BANKING

It was sincerely observed that consumer satisfaction is largely correlated with convenience (Aliyu, Rosmain, & Takala, 2014), reliability (Chong, Ooi, Lin, & Tan, 2010), responsiveness (Dhurup, Surujlal, & Redda, 2014), tangibility (Shompa & Islam, 2015), trust (Mukherjee & Nath, 2003), security (Damghanian, Zarei, & SiahsaraniKojuri, 2016) interface design quality (Shankar & Jebarajakirthy, 2019). These independent constructs have a very significant relationship with consumer satisfaction in the domain of digital banking.

4.4.4.4 CO-OCCURRENCE ANALYSIS OF DIFFERENT ANTECEDENTS AND CONSEQUENTS

A co-occurrence network is a collection of components that share a common location in a piece of writing. A network is formed by connecting factors based on a set of criteria that determine co-occurrence. Several studies on service quality and customer satisfaction are shown in this picture (Figure

4.1). Service quality and customer satisfaction may be measured using the numerous SERVQUAL dimensions. Also, we can observe how many research papers relate to and develop the relationship between each other through text-mining tools. Internet banking's compatibility, usage, awareness, and accessibility have all been shown in many studies within our scope to be substantially linked. Demographic characteristics like age and gender correlate with transactional access, too.

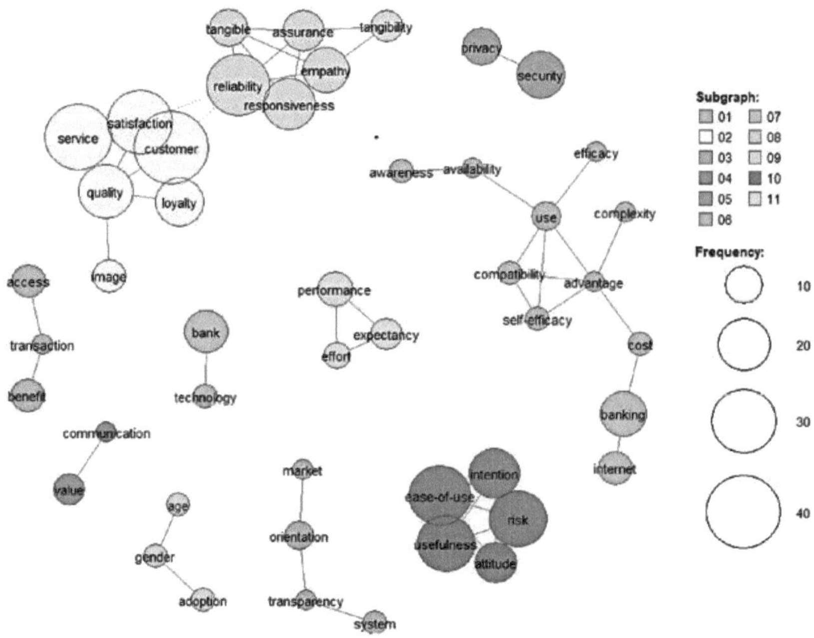

FIGURE 4.1 Co-occurrence analysis of different antecedents and consequents.

Source: Authors.

4.4.5 KEYWORDS USED IN ONLINE BANKING

A keyword is a code that unlocks information. Keywords direct researchers to relevant publications that would not otherwise come to the researcher's notice over the course of his reading (Joshi, 2014). Relevant articles may be unnoticed because they are published in journals that a certain researcher does not read on a regular basis (Rodrigues, 2013). Even if such articles are published in journals that the researcher reads on a regular basis, he or she may be unaware that they are important since their titles fail to signal their significance (Rodrigues, 2013).

4.4.5.1 CO-OCCURRENCE ANALYSIS OF DIFFERENT KEYWORDS OF THE REVIEWED JOURNAL

A co-occurrence network is a collection of components that appear in a text together. A network connects items especially keywords based on co-occurrence criteria. Figure 4.2 shows studies on different keywords – their usage, their frequency of usage and concentration. Text mining tools also enable us to see how many research publications are related to one another. Many studies in our field have shown a clear correlation between different elements of online banking.

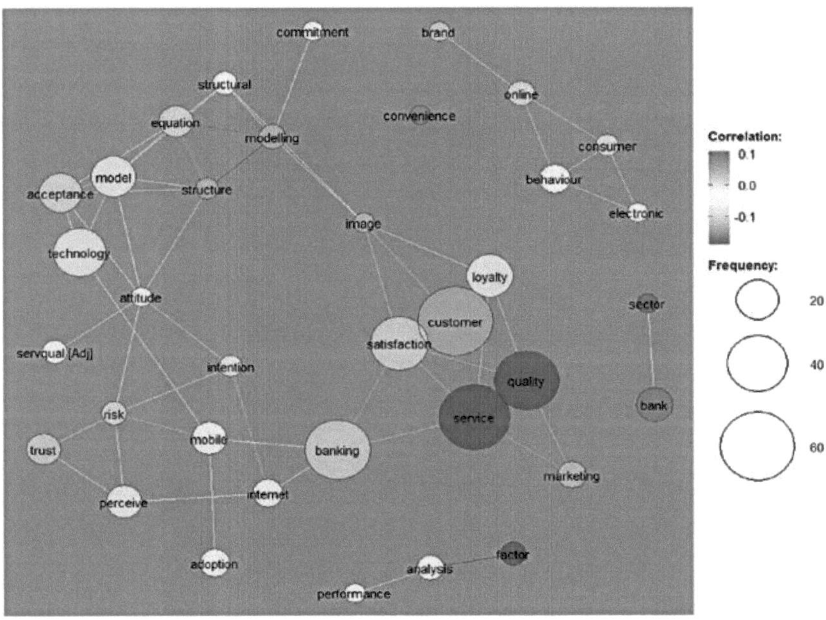

FIGURE 4.2 Co-occurrence analysis of different keywords of the reviewed journal.
Source: Authors.

4.4.5.2 MULTI-DIMENSIONAL ANALYSIS OF DIFFERENT KEYWORDS OF THE REVIEW JOURNAL

The multidimensional analysis examines things with meaningful hierarchies. This allows users to look at data from several perspectives. Here in Figure 4.3, the authors have tried to unveil the most frequently used keywords and their relationship in different articles. Using this method, the authors were able to compute the significant level according to their frequency of keywords and divide them into different clusters.

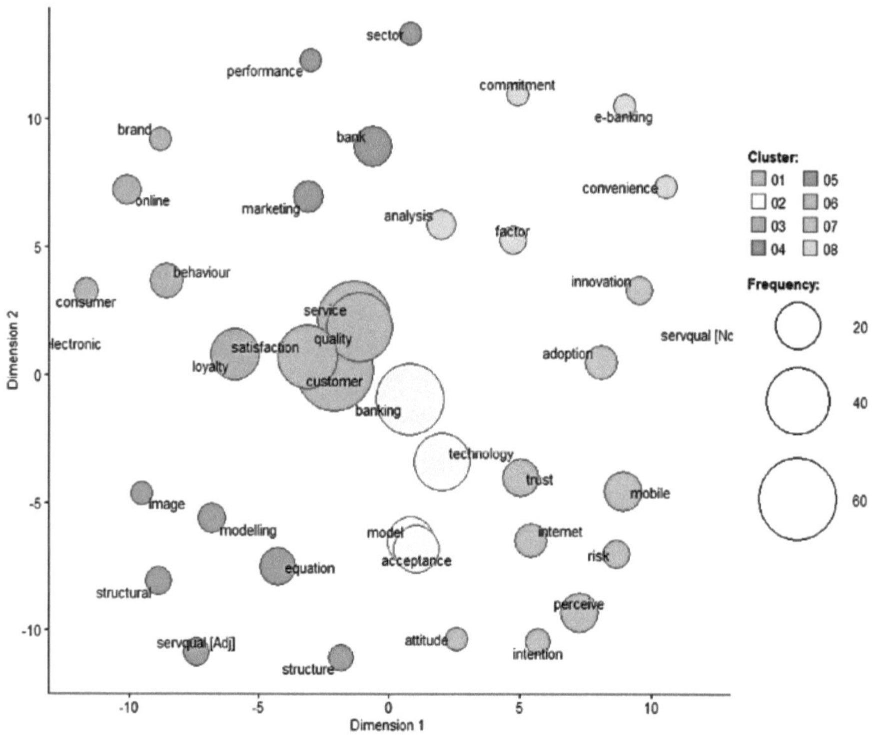

FIGURE 4.3 Multi-dimensional analysis of different keywords of the review journal.

Source: Authors.

4.5 CONCLUSIONS

To better understand the fundamental concepts of online digital banking, this chapter synthesizes and reviews the findings of 158 Scopus-indexed articles published between 1997 and 2021. It then proposes new research directions using the identified antecedents to further the body of knowledge on online digital banking. This study is based on the principles of systematic reviews. Starting with a list of domain-specific publications, we searched for those that had been indexed or were freely available. We have relied on journal indexing for the majority of our screening, and articles that are included in Scopus, Web of Science, and ABDC are given top priority. To narrow down the results after indexing, we looked at the articles' accessibility. The journals that may be accessed are arranged alphabetically by year of publication. Second, the papers are arranged by idea domain, i.e., Zoner, so that they are easily accessible. After that, we made an effort to organize the articles according to their approach, tried to extract the most important antecedents, consequents, and keywords, and then tried to decode them according to their frequency of usage and correlation amongst each other through text mining software tools. We're using this strategy to see if we can find any research gaps that could help us in the future.

The authors here used text mining algorithms to discover correlations between different constructs and their relative use in other publications. The writers wanted to find the optimum factors and techniques. For the study, they used co-occurrence and multidimensional cluster analysis to group the antecedents and consequents and the keywords to establish logical co-relation. In the online digital banking system, this review-based study gives clients an accurate image of the best feasible structures that have a large effect.

4.6 IMPLICATIONS

This study helps bankers to discover the primary essential aspects of online digital banking services to assess consumers' expectations and satisfaction. This chapter may assist online banking strategists and policymakers discover crucial aspects with which to work and aid their consumers by delivering or adopting cutting-edge, robust, convenient, and user-friendly technology. This study not only helps the bankers but also academically the researchers may gain the ingredients of their future research by identifying the proper antecedents and consequents.

4.7 LIMITATIONS AND SCOPE OF FUTURE RESEARCH

From the standpoint of research, such a review time may seem to be exceedingly selective. It is quite possible that important research published before 1997 was removed. Furthermore, the papers for this study were only gathered from Scopus, limiting the number of studies accessible for review. Despite these restrictions, the insights presented in this chapter will be useful to bankers, researchers, and academics all around the globe. Given the rising digitalization of marketing, it is critical that future research investigate consumers' perceptions of banking services while accounting for the impact of new-age innovative media. Additionally, the authors here only consider a qualitative SLR, and they are not reviewing the articles in the bibliometric and meta-analytical quantitative perspective.

Customers' post-adoption experiences and perspectives on digital banking are intriguing to study subjects. We've seen a lot of work with various models (Jebarajakirthy & Shankar, 2021; Ahmad, Bhatti, & Hwang, 2020), but very few works combine multiple contextual models with mixed and optimal approaches to reflect consumers' behavioral intention and satisfaction. In a growing nation like India, understanding how demographic and cultural factors affected the banking digital transformation process is critical. In the event of future research scope, it is also vital to explore how the digital divide (from the viewpoint of consumers) influences the banking digitization process in the context of demographic and psychographic antecedents.

KEYWORDS

- **antecedents**
- **consequents**
- **consumer loyalty**
- **consumer satisfaction**
- **digital banking**
- **online banking**
- **text mining**

REFERENCES

Aduda, J., & Kingoo, N., (2012). The relationship between electronic banking and financial performance among commercial banks in Kenya. *Journal of Finance and Investment Analysis*, *1*(3), 99–118.

Ahmad, N., Naveed, A., Ahmad, S., & Butt, I., (2020). Banking sector performance, profitability, and efficiency: A citation-based systematic literature review. *Journal of Economic Surveys*, *34*(1), 185–218.

Ahmad, S., Bhatti, S. H., & Hwang, Y., (2020). E-service quality and actual use of e-banking: Explanation through the technology acceptance model. *Information Development*, *36*(4), 503–519. https://doi.org/10.1177/0266666919871611.

Akhtar, S., Irfan, M., Sarwar, A., Asma, & Rashid, Q. U. A., (2019). Factors influencing individuals' intention to adopt mobile banking in China and Pakistan: The moderating role of cultural values. *Journal of Public Affairs*, *19*(1). https://doi.org/10.1002/pa.1884.

Alalwan, A. A., Dwivedi, Y. K., & Rana, N. P., (2017). Factors influencing adoption of mobile banking by Jordanian bank customers: Extending UTAUT2 with trust. *International Journal of Information Management*, *37*(3), 99–110. https://doi.org/10.1016/j.ijinfomgt.2017.01.002.

Aliyu, A. A., Rosmain, T., & Takala, J., (2014). Online banking and customer service delivery in Malaysia: Data screening and preliminary findings. *Procedia-Social and Behavioral Sciences*, *129*, 562–570.

Alizadeh, M., Greenberg, A., Maltz, D. A., Padhye, J., Patel, P., Prabhakar, B., & Sridharan, M., (2010). Data center TCP (DCTCP). In: *Proceedings of the ACM SIGCOMM 2010 Conference* (pp. 63–74).

Alkhowaiter, W. A., (2020). Digital payment and banking adoption research in Gulf countries: A systematic literature review. *International Journal of Information Management*, *53*, 102102.

Allada, V. K., & Dubey, R., (2014). Investigating the online banking customer satisfaction model under the controlled effect of demographic variables. *International Journal of Services and Operations Management*, *18*(3), 281–304.

Arbore, A., & Busacca, B., (2009). Customer satisfaction and dissatisfaction in retail banking: Exploring the asymmetric impact of attribute performances. *Journal of Retailing and Consumer Services*, *16*(4), 271–280.

Astrachan, C. B., Patel, V. K., & Wanzenried, G., (2014). A comparative study of CB-SEM and PLS-SEM for theory development in family firm research. *Journal of Family Business Strategy*, *5*(1), 116–128.

Asubonteng, P., McCleary, K. J., & Swan, J. E., (1996). SERVQUAL revisited: A critical review of service quality. *Journal of Services Marketing*.

Babakus, E., & Boller, G. W., (1992). An empirical assessment of the SERVQUAL scale. *Journal of Business Research*, *24*(3), 253–268.

Bowen, J. T., & Chen, S. L., (2001). The relationship between customer loyalty and customer satisfaction. *International Journal of Contemporary Hospitality Management*.

Chong, A. Y. L., Ooi, K. B., Lin, B., & Tan, B. I., (2010). Online banking adoption: An empirical analysis. *International Journal of Bank Marketing*.

Cooil, B., Keiningham, T. L., Aksoy, L., & Hsu, M., (2007). A longitudinal analysis of customer satisfaction and share of wallet: Investigating the moderating effect of customer characteristics. *Journal of Marketing*, *71*(1), 67–83.

Damghanian, H., Zarei, A., & SiahsaraniKojuri, M. A., (2016). Impact of perceived security on trust, perceived risk, and acceptance of online banking in Iran. *Journal of Internet Commerce*, *15*(3), 214–238.

De Oliveira, M. F. F., & Hwang, Y., (2018). An empirical investigation on disclosure about mobile banking on bank websites. *Online Information Review*, *42*(5), 615–629. https://doi.org/10.1108/OIR-05-2016-0136.

DeLone, W. H., & McLean, E. R., (1992). Information systems success: The quest for the dependent variable. *Information Systems Research*, *3*(1), 60–95.

Demirgüç-Kunt, A., & Detragiache, E., (2000). Monitoring banking sector fragility: A multivariate logit approach. *The World Bank Economic Review, 14*(2), 287–307.

Dhurup, M., Surujlal, J., & Redda, E., (2014). Customer perceptions of online banking service quality. *Mediterranean Journal of Social Sciences, 5*(2), 587–594.

Dinçer, H., Yüksel, S., Eti, S., & Tula, A., (2019). Effects of demographic characteristics on business success: An evidence from Turkish banking sector. In: *Handbook of Research on Business Models in Modern Competitive Scenarios* (pp. 304–324). IGI Global.

Diop, E. B., Zhao, S., & Duy, T. V., (2019). An extension of the technology acceptance model for understanding travelers' adoption of variable message signs. *PLoS One, 14*(4). https://doi.org/10.1371/journal.pone.0216007.

Durkin, M., O'Donnell, A., Mullholland, G., & Crowe, J., (2007). On e–banking adoption: From banker perception to customer reality. *Journal of Strategic Marketing, 15*(2, 3), 237–252. https://doi.org/10.1080/09652540701318815.

Goudarzi, S., (2013). *Customer Trust on Internet Banking Adoption.* Doctoral dissertation, Universiti Teknologi Malaysia.

Guriting, P., & Ndubisi, N. O., (2006). Borneo online banking: Evaluating customer perceptions and behavioral intention. *Management Research News.*

Hinson, R. E., (2011). Banking the poor: The role of mobiles. *Journal of Financial Services Marketing, 15*(4), 320–333.

Iberahim, H., Taufik, N. M., Adzmir, A. M., & Saharuddin, H., (2016). Customer satisfaction on reliability and responsiveness of self-service technology for retail banking services. *Procedia Economics and Finance, 37*, 13–20.

Ibrahim, E. E., Joseph, M., & Ibeh, K. I., (2006). Customers' perception of electronic service delivery in the UK retail banking sector. *International Journal of Bank Marketing.*

Ismagilova, E., Slade, E. L., Rana, N. P., & Dwivedi, Y. K., (2019). The effect of electronic word of mouth communications on intention to buy: A meta-analysis. *Information Systems Frontiers*, 1–24.

Jayachandran, A., (2019). E-Banking or branch banking ? preference of senior citizens in Kerala. *The IUP Journal of Bank Management, 18*(2), 19–29.

Jebarajakirthy, C., & Shankar, A., (2021). Impact of online convenience on mobile banking adoption intention: A moderated mediation approach. *Journal of Retailing and Consumer Services, 58*, 102323.

Joshi, Y., (2014). *Why Do Journals Ask for Keywords?* Web site: http://www.editage.com/insights/why-do-journals-ask-for-keywords (accessed on 8 December 2023).

Kaura, V., (2013). Antecedents of customer satisfaction: A study of Indian public and private sector banks. *International Journal of Bank Marketing, 31*(3), 167–186.

Krishnamoorthy, B., & Shivkumar, A., (2020). Punjab national bank: Implementing core banking solution. *International Journal of Services, Technology and Management, 26*(6), 502–519. https://doi.org/10.1504/IJSTM.2020.110366.

Kumar, A., Dhingra, S., Batra, V., & Purohit, H., (2020). A Framework of mobile banking adoption in India. *Journal of Open Innovation: Technology, Market, and Complexity, 6*(2). https://doi.org/10.3390/JOITMC6020040.

Kurmanova, D., Ajupov, A., Abelguzin, N., & Kurmanova, L., (2016). A comprehensive assessment of the financial security of banking system. *International Business Management, 10*(23), 5620–5624.

Ladhari, R., (2009). A review of twenty years of SERVQUAL research. *International Journal of Quality and Service Sciences.*

Ladhari, R., (2010). Developing e-service quality scales: A literature review. *Journal of Retailing and Consumer Services*, *17*(6), 464–477.

Lay, H. T., Boon, C. C., & Syaiful, R. H., (2016). Service innovation in Malaysian banking industry towards sustainable competitive advantage through environmentally and socially practices. *Procedia – Social and Behavioral Sciences*, *224*, 52–59. https://doi.org/10.1016/j.sbspro.2016.05.399.

Li, F., Lu, H., Hou, M., Cui, K., & Darbandi, M., (2021). Customer satisfaction with bank services: The role of cloud services, security, e-learning, and service quality. *Technology in Society*, *64*. https://doi.org/10.1016/j.techsoc.2020.101487.

Mahamad, O., & Ramayah, T., (2010). Service quality, customer satisfaction and loyalty: A test of mediation. *International Business Research*, *3*(4), 72.

Mano, H., & Oliver, R. L., (1993). Assessing the dimensionality and structure of the consumption experience: Evaluation, feeling, and satisfaction. *Journal of Consumer Research*, *20*(3), 451–466.

Mansour, K. B., Kooli, K., & Utama, R., (2014). Online trust antecedents and their consequences on purchase intention: An integrative approach. *Journal of Customer Behavior*, *13*(1), 25–42.

McKnight, D. H., & Chervany, N. L., (2001). What trust means in e-commerce customer relationships: An interdisciplinary conceptual typology. *International Journal of Electronic Commerce*, *6*(2), 35–59.

Mehrad, D., & Mohammadi, S., (2017). Word of mouth impact on the adoption of mobile banking in Iran. *Telematics and Informatics*, *34*(7), 1351–1363.

Mukherjee, A., & Nath, P., (2003). A model of trust in online relationship banking. *International Journal of Bank Marketing*.

Muñoz-Leiva, F., Climent-Climent, S., & Liébana-Cabanillas, F., (2017). Determinants of intention to use the mobile banking apps: An extension of the classic TAM model. *Spanish Journal of Marketing – ESIC*, *21*(1), 25–38.

Oliver, R. L., (1981). Measurement and evaluation of satisfaction processes in retail settings. *Journal of Retailing*.

Prabowo, H., (2016). Learning fraud detection from big data in online banking transactions: A systematic literature review. *Journal of Telecommunication, Electronic and Computer Engineering (JTEC)*, *8*(3), 127–131.

Pradhan, K., Ghule, G. S., Yadav, D. R., & Shinde, S. S., (2020). Banking system using block chain technology. *International Journal of Scientific Research in Science, Engineering and Technology*, 566–570. https://doi.org/10.32628/ijsrset2073112.

Rahayu, M., Rasid, F., & Tannady, H., (2018). Effects of self-efficacy, job satisfaction, and work culture toward performance of telemarketing staff in banking sector. *South East Asia J. Contemp. Business, Econ. Law*, *16*(5), 47–52.

Rajan, A. P., & Shamini, D. A. R., (2018). Information technology plays a vital role in banking sector: Future banking services. *Journal of Emerging Technologies and Innovative Research*, *5*(11), 414–417.

Rastogi, S., Panse, C., Sharma, A., & Bhimavarapu, V. M., (2021). Unified payment interface (UPI): A digital innovation and its impact on financial inclusion and economic development. *Universal Journal of Accounting and Finance*, *9*(3), 518–530. https://doi.org/10.13189/ujaf.2021.090326.

Rhaiem, K., & Amara, N., (2019). Learning from innovation failures: A systematic review of the literature and research agenda. *Review of Managerial Science*, 1–46.

Rodrigues, V., (2013). *How to Write an Effective Title and Abstract and Choose Appropriate Keywords.* Editage Insights.

Seth, N., Deshmukh, S. G., & Vrat, P., (2005). Service quality models: A review. *International Journal of Quality & Reliability Management.*

Shahabi, V., & Faezy, R. F., (2019). Modeling the effect of electronic banking expansion on profitability using neural networks and system dynamics approach. *Qualitative Research in Financial Markets*, *11*(2), 197–210. https://doi.org/10.1108/QRFM-11-2018-0117.

Shankar, A., & Jebarajakirthy, C., (2019). The influence of e-banking service quality on customer loyalty: A moderated mediation approach. *International Journal of Bank Marketing.*

Shergill, G. S., & Li, B., (2005). Internet banking-an empirical investigation of a trust and loyalty model for New Zealand banks. *Journal of Internet Commerce, 4*(4), 101–118.

Shompa, F., & Islam, Z., (2015). Customer satisfaction of online banking services: A case study on Islamic banks in Bangladesh. *Bangladesh Research Journal, 15.*

Teeroovengadum, V., (2020). Service quality dimensions as predictors of customer satisfaction and loyalty in the banking industry: Moderating effects of gender. *European Business Review.*

Thakur, R., (2014). What keeps mobile banking customers loyal? *International Journal of Bank Marketing, 32*(7), 628–646.

Thorpe, R., Holt, R., Macpherson, A., & Pittaway, L., (2005). Using knowledge within small and medium-sized firms: A systematic review of the evidence. *International Journal of Management Reviews*, *7*(4), 257–281.

Tong, C., Wong, S. K. S., & Lui, K. P. H., (2012). The influences of service personalization, customer satisfaction and switching costs on e-loyalty. *International Journal of Economics and Finance, 4*(3), 105–114.

Unyathanakorn, K., Kasikornbank, P. C. L., & Rompho, N., (2014). Factors affecting customer satisfaction in online banking service. *Journal of Marketing Development and Competitiveness, 8*(2), 50.

Wu, P., (2017). Fintech trends relationships research: A bibliometric citation meta-analysis. In: *Proceedings of the 17th International Conference on Electronic Business, ICEB, Dubai, UAE* (pp. 99–105).

Yadav, D. R., Ghule, G. S., & Shinde, S. S., (2020). Survey on banking system using block chain technology. *International Journal of Scientific Research in Science, Engineering and Technology*, 368–371. https://doi.org/10.32628/ijsrset196635.

Yoon, C., (2010). Antecedents of customer satisfaction with online banking in China: The effects of experience. *Computers in Human Behavior, 26*(6), 1296–1304.

Zhu, X., Yan, R., Yang, J., & Xu, T., (2021). Online promotion of mobile banking activities - a comparative study of the US and Chinese banks. *International Journal of Mobile Communications*, *19*(4), 439. https://doi.org/10.1504/ijmc.2021.116118.

CHAPTER 5

An Empirical Study to Understand the Parameters Affecting Customers' Satisfaction Toward Fashion Brands Following Sustainable Practices

VARSHA MISHRA,[1] JYOTI SHAW,[2] ANKITA SINGH,[2] ADITYA VIKRAM SINGH,[2] and HENA IQBAL[3]

[1]*MSc Digital Marketing, Royal Holloway, University of London, UK*

[2]*Jagdish Sheth School of Management(JAGSoM), Bangalore, India*

[3]*Adjunct Faculty, Murdoch University, Dubai*

ABSTRACT

With the passage of time and continuous evolution in the market, the necessity of clothing has become a symbol of fashion in society. The change in customer fashion taste has brought about a revolution and upsurge in the fashion industry. Brands are striving to offer innovative concepts for the welfare of society, in addition to catering to the interests of their clients. However, modern fashion is wreaking havoc on the environment by causing a maximum impact on nature. The waste material from the clothing industry dumps into the drainage that eventually connects to the rivers. This wastage can be treacherous for both marine and human life. With an increase in awareness among the people, there has been an adoption of sustainable production and consumption worldwide. As a result, many fashion firms are shifting to sustainable practices throughout their manufacturing processes as part of their corporate social responsibility to continue gaining loyal customers who are environmentally conscious.

Sustainability in Marketing Practice: Strategies for Industry 4.0.
Subrata Chattopadhyay, Udit Chawla, (Eds.)

This research work aims to evaluate customer satisfaction with fashion brands adopting sustainable practices throughout their manufacturing process. We have considered different brands, namely Brand-A, Brand-B, Brand-C, Brand-D, Brand-E, and Brand-F, along with five different parameters like price, quality, variety, color, and fittings, concerning casual wear brands following sustainable practices.

With the help of ANOVA analysis, a significant association has been found between the parameters of price, quality, and variety with the different casual wear brands following sustainable practices. Customers are found to be satisfied with the brands as they have maintained these parameters with an environmental conservation policy. Hence, it is implied that these parameters play a crucial part in determining consumer satisfaction concerning casual wear fashion and environmental protection.

5.1 INTRODUCTION

In past decades, sustainability has evolved from a market trend to a necessary product. Earlier, it was considered a marketing gimmick to attract customers (Jung et al., 2020). Over time, industries have understood the importance that ecological products possess. With the fourth revolution of industry, the usage of giant machinery and non-renewable resources has increased significantly to meet the requirements of the growing population. Such machinery emits an immense amount of carbon emissions and other types of pollution, which can cause harmful effects on the environment. The rise in industrial emissions into the atmosphere can be toxic to human existence. The constant peak in pollution and its consequences on human life has created a demand for environmentally-friendly and sustainable goods. Sustainable marketing is the framework for any brand's social and economic practices (Chabowski et al., 2011; Elkington, 1998). Sustainable marketing is a marketing strategy to influence the behavior of individuals to consume, as well as the way the business manufactures, finding the correct balance between the two will lead to enormously significant societal and natural benefits. Industries that practice sustainability consider it a social responsibility to conserve the environment without prioritizing high-profit margins on the product (Ducrot & Murat, 2011). Modern-day studies suggest sustainable services and products as a necessity for human civilization (Dewanto & Belgaiwan, 2020).

Clothing or garments are one of the oldest and biggest existing industries. According to reports, the capital market for the global fashion industry is $3 billion, which contributes around 2% of global GDP (Wang et al., 2017). It is the highest-grossing industry of all time. Clothing is considered a necessity for livelihood. With the evolution of human culture, clothing has become more of a status symbol than a necessity. In recent years, the clothing industry has expanded significantly to meet the requirements of the growing population. The expansion of the garment industry causes rising pollution in the environment on a mass scale (Gam & Banning, 2011). The clothing sector accounts for about 10% of global carbon emissions. It can be cited as the industry that has the maximum impact on nature. The fashion industry is dependent on natural derivatives for its production. From fabrication to land, the fashion industry relies on the environment. Hence, it relies on a large sector of natural resources, generating waste materials that pollute the natural resources on a mass scale (Fletcher, 2013). The conversion of natural materials to end items has detrimental outcomes on the environment. Therefore, textile industries are required to incorporate environmentally-friendly activities to minimize detrimental effects as much as possible. According to Shen et al. (2012), there is evidence that customers, aware of the detrimental effects caused by textile industries, prefer and purchase apparel from brands that follow sustainable practices, even if they have to pay extra for them.

The trend of sustainable products was visible after the sudden shift in customer demand for sustainable products (Mittelstaedt et al., 2014; Chabowski et al., 2011; Bendell & Kleanthous, 2007; Elkington, 1998). With the rise in awareness among the customers, the market for green goods is growing. The marketplace for this segment is increasing with the passage of time. Industries have caught it as an opportunity to capture the market for sustainable products (Nidumolu et al., 2009; Pfeffer, 2010; Sheth et al., 2011), helping fashion appear as the primary sustainable industry (Watson & Yan, 2013; Mora et al., 2014). As per some studies present in the literature, adoption of sustainable activities by the textile companies attracted a large base of environmentally concerned customers (Shen et al., 2014; Ko et al., 2013). The increased emphasis on sustainable clothes has prompted clothing firms and merchants to respond (Niinimäki & Hassi, 2011). Many fashion firms are beginning to include sustainability into their manufacturing processes as part of their corporate social responsibility and making sure consumers perceive it (Puspita & Chae, 2020) because the consumer is mostly likely to boycott irresponsible companies. By adopting sustainable methods, businesses can attract more consumers and achieve their commercial goals while also contributing to society (Di Benedetto, 2017; Mohr et al., 2001). As a result of changing consumer habits, the fashion industry is becoming more environmentally friendly (Niinimäki & Hassi, 2011). Sustainable marketing helps the brand to create a positive image brand in a customer perspective and leads to corporate growth (Martínez & Del Bosque, 2013; Prates et al., 2015). Consequently, textile company's environmentally friendly operations foster a strong company perception and attitude toward consumers; as a result, an organization earns a competitive edge over others including an upsurge in equity (Kandampully & Suhartanto, 2003; Porter & Kramer, 2006).

5.2 REVIEW OF LITERATURE

Sustainable or environmentally friendly marketing is a newer concept introduced in the fashion industry after consumers tended toward the environment (Jung & Jin, 2014). The sustainable fashion industry describes the industry that pays attention to the present customer need without causing harm to the surroundings for the future generation (UNWCED, 1987). According to Elkington (1997), sustainable marketing operations in the fashion industry should have three components such as economic, social, and environmental. Sustainable marketing in the fashion industry is an umbrella term for marketing practices aimed to cause minimal detrimental effects on the environment throughout the process of producing and selling textile products (Kim et al., 2015; Kim & Schellhase, 2015; Sun & Ko, 2016). Sustainable fashion comprises a mix of concepts focusing on the environment, animal exploitation in the textile industry, sustainable consumption, forced labor, and many others (Cervellon et al., 2010; Bray, 2009; Bianchi & Birtwistle, 2010; Blanchard, 2013).

In a sustainable fashion market, price is one of the essential parameters to discuss. Customers' buying behavior can be determined by their intention to purchase the product. A customer will waver to pay more if the product is not up to the mark and will halt their purchase decision (Chan & Wong, 2012). It becomes crucial for a company to have fair pricing on its products as a part of its corporate social responsibility. Brands offering appropriate pricing will attract new customers and develop a sense of goodwill among the existing customers (Dekimpe & Hanssens, 2018). It arises as a challenge for brands to maintain the quality and supply chain of the product at such an inferior cost as sustainable practices are expensive methods (Shen et al., 2017). From a brand perspective, it is necessary to maintain symmetry between its profit margin and its responsibility toward the societal benefit (Chang et al., 2017; Mazanov & Woolf, 2017).

The purchase intention of the buyer depends on the brand motive behind the sustainable offering. A customer will accept any product from a particular brand if they successfully deliver a quality product with an environmental conservation reason behind (Chan & Wong, 2012). Textile brands have implemented a green building strategy by recycling waste materials throughout the supply chain of the product to overcome the rising drainage and waste (Ciasullo et al., 2017). Such practices by brands will build a positive image for the brand in society. Fashion product designs frequently incorporate green and recycled raw materials to increase consumer happiness, although fashion consumers judge fashion items mostly on aesthetics (Howarth & Hadfield, 2006; Joergens, 2008).

The quality of the product matters the most in the demand creation of the product (Niinimäki, 2010). The consumer will show their commitment to the brand if they triumphantly deliver a fine product. According to our study, the most influential aspect of a purchase decision for any fashion customer depends on the price, quality, and the variety of options available. Customers will pay if brands don't compromise the quality of the product. Such initiatives by brands will create a sense of loyalty among the customers. Customers and organizations benefit from socially beneficial co-creation initiatives when consumers are dedicated to businesses and invest effort into producing value. As a result, customers get more emotionally engaged with the items. Upmarket apparels make customers feel like higher-level people in the community, thereby inducing the habit of frequent purchase among them (Niinimäki, 2010). By this method, some brands display their corporate social responsibility toward society and the environment along with generating profit (Wang et al., 2017; Knox & Maklan, 2004).

Customers choosing sustainable fashion products are limited to absolute numbers of products. There is very little chance of having a variety of sustainable fashion products. Joergens (2006) stated that the customers get limited alternatives for sustainable clothing, due to their incomparable cost, as compared to other fashion alternatives open to them. She observed that the customers find the design and presentation of sustainable clothes quite unappealing and mismatching to personal needs. Additionally, customers claim that while choosing clothes, features such as cost, standard, and appearance will supersede customer ideals; clothes should be both environmentally friendly and attractive to the consumer's aesthetic standards (Beard, 2008).

5.3 OBJECTIVE

The objective behind the study is to identify whether there is any significant difference between the parameters of sustainable marketing in the fashion industry.

5.4 RESEARCH METHODOLOGY

This research work is aimed to evaluate customer satisfaction with fashion brands adopting sustainable practices throughout their manufacturing process. We have considered different brands, that is Brand-A, Brand-B, Brand-C, Brand-D, Brand-E, and Brand-F along with five different parameters like price, fabric, quality, variety, and fittings, concerning casual wear brands following sustainable practices. Responses were collected from Kolkata, West Bengal by incorporating a structured questionnaire. Kolkata is the capital city of West Bengal and one of the most popular metropolitan cities. Throughout history, Kolkata is known for its mills and clothing industry in East India. The sustentation habits of people and the growing amount of sustainability knowledge is the reason for Kolkata as a data collection center.

The data is encapsulated with the help of SPSS software to attain the objective of the study. With the aid of ANOVA analysis, significant associations have been found between the parameters, price, quality, and variety with the different casual wear brands following sustainable practices. Customers are found to be satisfied with the brands as they have maintained these parameters with environmental conservation policy. Hence, it is implied that these parameters play a crucial part in determining consumer satisfaction concerning casual wear fashion and environmental protection.

5.5 FINDINGS AND ANALYSIS

5.5.1 HYPOTHESES

We have considered two hypotheses for our study which are:

- **H_0:** Implies Customer Satisfaction do not possess any substantial relationship with the parameters of Sustainable Marketing in the Fashion Industry.
- **H_1:** Implies Customer Satisfaction possess substantial relationship with the parameters of Sustainable Marketing in the Fashion Industry.

From Table 5.1, we can see that the asymp. sig for *Price*, *Quality*, and *Variety* is less than 0.05, therefore accepting alternative hypothesis, in these cases. In addition, it implies a substantial difference between the parameters *Price*, *Quality*, and *Variety* concerning different brand preferences. To understand the difference between brands in-depth, Post Hoc Test (Tables 5.2(a)–(c)) was undertaken.

TABLE 5.1 Results of ANOVA

		Sum of Squares	df	Mean Square	F	Sig.
Color	Between groups	2686.96	5	537.392	1.916	0.092
	Within groups	65921.455	235	280.517	–	–
	Total	68608.415	240			
Price	Between groups	1793.916	5	358.783	2.405	0.038
	Within groups	35052.706	235	149.16	–	–
	Total	36846.622	240			
Fittings	Between groups	1190.542	5	238.108	0.674	0.644
	Within groups	83049.325	235	353.401	–	–
	Total	84239.867	240			
Quality	Between groups	2102.76	5	420.552	2.48	0.033
	Within groups	39847.282	235	169.563	–	–
	Total	41950.041	240			
Variety	Between groups	8160.144	5	1632.029	3.998	0.002
	Within groups	95934.894	235	408.234	–	–
	Total	104095.04	240			

Source: Primary Data Analysis Using SPSS 21.0.

From Table 5.2(a) of Post Hoc, Brand-B, Brand-D, and Brand-F are different than Brand-A, Brand-C, and Brand-E in terms of the price of the casuals. Table 5.2(a) implies that all the brands practicing sustainable methods are comparatively on the higher side of the cost of production than the regular production methods. Because sustainable methods of production are different from the regular production methods. It requires different types of machinery, materials, methods, supply chain, proper waste disposal, and utilization of resources. These methods and practices are relevantly more expensive than the traditional ones. Different brands charge different costs to their customer to meet their profit requirements.

TABLE 5.2(A) Results of Post Hoc Analysis for the Parameter "Price," of Sustainable Marketing in the Fashion Industry

Price	Brand A	Brand B	Brand C	Brand D	Brand E	Brand F
Brand A	–	H_0 accepted	H_0 accepted	H_0 accepted	H_0 accepted	H_0 accepted
Brand B	H_1 accepted	–	H_1 accepted	H_0 accepted	H_1 accepted	H_1 accepted
Brand C	H_0 accepted	H_0 accepted	–	H_0 accepted	H_0 accepted	H_0 accepted
Brand D	H_1 accepted	H_1 accepted	H_1 accepted	–	H_1 accepted	H_0 accepted
Brand E	H_0 accepted	H_0 accepted	H_0 accepted	H_0 accepted	–	H_0 accepted
Brand F	H_1 accepted	H_0 accepted	H_1 accepted	H_1 accepted	H_1 accepted	–

Source: Primary Data Analysis Using SPSS 21.0.

From Table 5.2(b) of Post Hoc, Brand-A, Brand-E, and Brand-F are different from the brands Brand-B, Brand-C, and Brand-D. It highlights the difference in the quality of products from these brands. Brand-A, Brand-E, and Brand-F offer a better quality of products to their customers. These brands follow their corporate social responsibility by providing better quality products that create less or no impact on the environment, as compared to traditional brands. Apart from using better quality fabric for production, this brands also offer a range of quality materials and products which attract more satisfied customers.

TABLE 5.2(B) Results of Post Hoc Analysis for the Parameter "Quality," of Sustainable Marketing in the Fashion Industry

Quality	Brand A	Brand B	Brand C	Brand D	Brand E	Brand F
Brand A	–	H_1 accepted	H_1 accepted	H_1 accepted	H_0 accepted	H_0 accepted
Brand B	H_0 accepted	–	H_1 accepted	H_1 accepted	H_0 accepted	H_0 accepted
Brand C	H_0 accepted	H_0 accepted	–	H_0 accepted	H_0 accepted	H_0 accepted
Brand D	H_0 accepted	H_0 accepted	H_0 accepted	–	H_0 accepted	H_0 accepted
Brand E	H_1 accepted	H_1 accepted	H_1 accepted	H_1 accepted	–	H_0 accepted
Brand F	H_1 accepted	H_1 accepted	H_1 accepted	H_1 accepted	H_0 accepted	–

Source: Primary Data Analysis Using SPSS 21.0.

From Table 5.2(c) of Post Hoc, Brand-A, Brand-D, Brand-E are different from the brands which are Brand-B, Brand-C, and Brand-F, that is they are doing better in terms of providing a variety of options to their customers. In the entire fashion industry, limited brands offer a range of products to their customers. It implies that these brands either possess limited resources or they face overutilization or underutilization of the resources. Brand-A, Brand-D, Brand-E successfully delivers a range of products to their customers because

TABLE 5.2(C) Results of Post Hoc Analysis for the Parameter "Variety," of Sustainable Marketing in the Fashion Industry

Variety	Brand A	Brand B	Brand C	Brand D	Brand E	Brand F
Brand A	–	H_1 accepted	H_1 accepted	H_1 accepted	H_1 accepted	H_0 accepted
Brand B	H_1 accepted	–	H_0 accepted	H_0 accepted	H_0 accepted	H_0 accepted
Brand C	H_1 accepted	H_0 accepted	–	H_0 accepted	H_0 accepted	H_0 accepted
Brand D	H_1 accepted	H_1 accepted	H_1 accepted	–	H_1 accepted	H_0 accepted
Brand E	H_1 accepted	H_1 accepted	H_1 accepted	H_0 accepted	–	H_1 accepted
Brand F	H_0 accepted	H_1 accepted	H_0 accepted	H_0 accepted	H_0 accepted	–

Source: Primary Data Analysis Using SPSS 21.0.

TABLE 5.3 Results of Multiple Comparisons

Dependent Variable	(I) Brands	(J) Brands	Mean Difference (I-J)	Std. Error	Sig.	95% Confidence Interval	
						Lower Bound	Upper Bound
Color	Brand A	Brand B	–7.945*	3.414	0.021	–14.67	–1.22
		Brand C	–9.220*	4.29	0.033	–17.67	–0.77
		Brand D	–10.194*	3.401	0.003	–16.89	–3.49
		Brand E	–8.4	5.296	0.114	–18.83	2.03
		Brand F	–5.368	6.26	0.392	–17.7	6.96
	Brand B	Brand A	7.945*	3.414	0.021	1.22	14.67
		Brand C	–1.275	3.746	0.734	–8.65	6.11
		Brand D	–2.248	2.682	0.403	–7.53	3.04
		Brand E	–0.455	4.866	0.926	–10.04	9.13
		Brand F	2.577	5.9	0.663	–9.05	14.2
	Brand C	Brand A	9.220*	4.29	0.033	0.77	17.67
		Brand B	1.275	3.746	0.734	–6.11	8.65
		Brand D	–0.974	3.734	0.794	–8.33	6.38
		Brand E	0.82	5.516	0.882	–10.05	11.69
		Brand F	3.852	6.447	0.551	–8.85	16.55
	Brand D	Brand A	10.194*	3.401	0.003	3.49	16.89
		Brand B	2.248	2.682	0.403	–3.04	7.53
		Brand C	0.974	3.734	0.794	–6.38	8.33
		Brand E	1.794	4.857	0.712	–7.77	11.36
		Brand F	4.826	5.892	0.414	–6.78	16.43
	Brand E	Brand A	8.4	5.296	0.114	–2.03	18.83
		Brand B	0.455	4.866	0.926	–9.13	10.04

TABLE 5.3 *(Continued)*

Dependent Variable	(I) Brands	(J) Brands	Mean Difference (I-J)	Std. Error	Sig.	95% Confidence Interval	
						Lower Bound	Upper Bound
		Brand C	–0.82	5.516	0.882	–11.69	10.05
		Brand D	–1.794	4.857	0.712	–11.36	7.77
		Brand F	3.032	7.156	0.672	–11.07	17.13
	Brand F	Brand A	5.368	6.26	0.392	–6.96	17.7
		Brand B	–2.577	5.9	0.663	–14.2	9.05
		Brand C	–3.852	6.447	0.551	–16.55	8.85
		Brand D	–4.826	5.892	0.414	–16.43	6.78
		Brand E	–3.032	7.156	0.672	–17.13	11.07
Price	Brand A	Brand B	–0.013	2.49	0.996	–4.92	4.89
		Brand C	1.228	3.128	0.695	–4.94	7.39
		Brand D	5.199*	2.48	0.037	0.31	10.08
		Brand E	1.429	3.862	0.712	–6.18	9.04
		Brand F	–5.698	4.565	0.213	–14.69	3.29
	Brand B	Brand A	0.013	2.49	0.996	–4.89	4.92
		Brand C	1.241	2.732	0.65	–4.14	6.62
		Brand D	5.212*	1.956	0.008	1.36	9.07
		Brand E	1.442	3.548	0.685	–5.55	8.43
		Brand F	–5.685	4.302	0.188	–14.16	2.79
	Brand C	Brand A	–1.228	3.128	0.695	–7.39	4.94
		Brand B	–1.241	2.732	0.65	–6.62	4.14
		Brand D	3.971	2.723	0.146	–1.39	9.34
		Brand E	0.201	4.022	0.96	–7.72	8.13

TABLE 5.3 *(Continued)*

Dependent Variable	(I) Brands	(J) Brands	Mean Difference (I-J)	Std. Error	Sig.	95% Confidence Interval	
						Lower Bound	Upper Bound
		Brand F	–6.926	4.701	0.142	–16.19	2.34
	Brand D	Brand A	–5.199*	2.48	0.037	–10.08	–0.31
		Brand B	–5.212*	1.956	0.008	–9.07	–1.36
		Brand C	–3.971	2.723	0.146	–9.34	1.39
		Brand E	–3.77	3.542	0.288	–10.75	3.21
		Brand F	–10.897*	4.297	0.012	–19.36	–2.43
	Brand E	Brand A	–1.429	3.862	0.712	–9.04	6.18
		Brand B	–1.442	3.548	0.685	–8.43	5.55
		Brand C	–0.201	4.022	0.96	–8.13	7.72
		Brand D	3.77	3.542	0.288	–3.21	10.75
		Brand F	–7.127	5.218	0.173	–17.41	3.15
	Brand F	Brand A	5.698	4.565	0.213	–3.29	14.69
		Brand B	5.685	4.302	0.188	–2.79	14.16
		Brand C	6.926	4.701	0.142	–2.34	16.19
		Brand D	10.897*	4.297	0.012	2.43	19.36
		Brand E	7.127	5.218	0.173	–3.15	17.41
Fittings	Brand A	Brand B	0.538	3.832	0.889	–7.01	8.09
		Brand C	0.274	4.815	0.955	–9.21	9.76
		Brand D	–3.559	3.817	0.352	–11.08	3.96
		Brand E	1.2	5.945	0.84	–10.51	12.91
		Brand F	–7.244	7.026	0.304	–21.09	6.6

TABLE 5.3 *(Continued)*

Dependent Variable	(I) Brands	(J) Brands	Mean Difference (I-J)	Std. Error	Sig.	95% Confidence Interval	
						Lower Bound	Upper Bound
	Brand B	Brand A	–0.538	3.832	0.889	–8.09	7.01
		Brand C	–0.264	4.205	0.95	–8.55	8.02
		Brand D	–4.097	3.01	0.175	–10.03	1.83
		Brand E	0.662	5.462	0.904	–10.1	11.42
		Brand F	–7.782	6.622	0.241	–20.83	5.26
	Brand C	Brand A	–0.274	4.815	0.955	–9.76	9.21
		Brand B	0.264	4.205	0.95	–8.02	8.55
		Brand D	–3.834	4.191	0.361	–12.09	4.42
		Brand E	0.926	6.191	0.881	–11.27	13.12
		Brand F	–7.519	7.236	0.3	–21.77	6.74
	Brand D	Brand A	3.559	3.817	0.352	–3.96	11.08
		Brand B	4.097	3.01	0.175	–1.83	10.03
		Brand C	3.834	4.191	0.361	–4.42	12.09
		Brand E	4.759	5.451	0.384	–5.98	15.5
		Brand F	–3.685	6.614	0.578	–16.71	9.34
	Brand E	Brand A	–1.2	5.945	0.84	–12.91	10.51
		Brand B	–0.662	5.462	0.904	–11.42	10.1
		Brand C	–0.926	6.191	0.881	–13.12	11.27
		Brand D	–4.759	5.451	0.384	–15.5	5.98
		Brand F	–8.444	8.032	0.294	–24.27	7.38
	Brand F	Brand A	7.244	7.026	0.304	–6.6	21.09
		Brand B	7.782	6.622	0.241	–5.26	20.83

TABLE 5.3 *(Continued)*

Dependent Variable	(I) Brands	(J) Brands	Mean Difference (I-J)	Std. Error	Sig.	95% Confidence Interval	
						Lower Bound	Upper Bound
		Brand C	7.519	7.236	0.3	–6.74	21.77
		Brand D	3.685	6.614	0.578	–9.34	16.71
		Brand E	8.444	8.032	0.294	–7.38	24.27
Quality	Brand A	Brand B	–6.003*	2.655	0.025	–11.23	–0.77
		Brand C	–5.79	3.335	0.084	–12.36	0.78
		Brand D	–2.09	2.644	0.43	–7.3	3.12
		Brand E	–9.529*	4.118	0.022	–17.64	–1.42
		Brand F	–11.013*	4.867	0.025	–20.6	–1.42
	Brand B	Brand A	6.003*	2.655	0.025	0.77	11.23
		Brand C	0.212	2.912	0.942	–5.53	5.95
		Brand D	3.913	2.085	0.062	–0.2	8.02
		Brand E	–3.526	3.783	0.352	–10.98	3.93
		Brand F	–5.01	4.587	0.276	–14.05	4.03
	Brand C	Brand A	5.79	3.335	0.084	–0.78	12.36
		Brand B	–0.212	2.912	0.942	–5.95	5.53
		Brand D	3.7	2.903	0.204	–2.02	9.42
		Brand E	–3.738	4.289	0.384	–12.19	4.71
		Brand F	–5.222	5.012	0.299	–15.1	4.65
	Brand D	Brand A	2.09	2.644	0.43	–3.12	7.3
		Brand B	–3.913	2.085	0.062	–8.02	0.2
		Brand C	–3.7	2.903	0.204	–9.42	2.02
		Brand E	–7.439	3.776	0.05	–14.88	0

TABLE 5.3 *(Continued)*

Dependent Variable	(I) Brands	(J) Brands	Mean Difference (I-J)	Std. Error	Sig.	95% Confidence Interval	
						Lower Bound	Upper Bound
		Brand F	–8.923	4.581	0.053	–17.95	0.1
	Brand E	Brand A	9.529*	4.118	0.022	1.42	17.64
		Brand B	3.526	3.783	0.352	–3.93	10.98
		Brand C	3.738	4.289	0.384	–4.71	12.19
		Brand D	7.439	3.776	0.05	0	14.88
		Brand F	–1.484	5.563	0.79	–12.44	9.48
	Brand F	Brand A	11.013*	4.867	0.025	1.42	20.6
		Brand B	5.01	4.587	0.276	–4.03	14.05
		Brand C	5.222	5.012	0.299	–4.65	15.1
		Brand D	8.923	4.581	0.053	–0.1	17.95
		Brand E	1.484	5.563	0.79	–9.48	12.44
Variety	Brand A	Brand B	–13.821*	4.119	0.001	–21.94	–5.71
		Brand C	–17.753*	5.175	0.001	–27.95	–7.56
		Brand D	–12.052*	4.103	0.004	–20.13	–3.97
		Brand E	–15.957*	6.389	0.013	–28.54	–3.37
		Brand F	–26.013*	7.551	0.001	–40.89	–11.14
	Brand B	Brand A	13.821*	4.119	0.001	5.71	21.94
		Brand C	–3.933	4.519	0.385	–12.84	4.97
		Brand D	1.769	3.236	0.585	–4.61	8.14
		Brand E	–2.136	5.87	0.716	–13.7	9.43
		Brand F	–12.192	7.118	0.088	–26.21	1.83
	Brand C	Brand A	17.753*	5.175	0.001	7.56	27.95

TABLE 5.3 *(Continued)*

Dependent Variable	(I) Brands	(J) Brands	Mean Difference (I-J)	Std. Error	Sig.	95% Confidence Interval	
						Lower Bound	Upper Bound
		Brand B	3.933	4.519	0.385	–4.97	12.84
		Brand D	5.701	4.504	0.207	–3.17	14.58
		Brand E	1.796	6.654	0.787	–11.31	14.91
		Brand F	–8.259	7.777	0.289	–23.58	7.06
	Brand D	Brand A	12.052*	4.103	0.004	3.97	20.13
		Brand B	–1.769	3.236	0.585	–8.14	4.61
		Brand C	–5.701	4.504	0.207	–14.58	3.17
		Brand E	–3.905	5.859	0.506	–15.45	7.64
		Brand F	–13.961	7.108	0.051	–27.96	0.04
	Brand E	Brand A	15.957*	6.389	0.013	3.37	28.54
		Brand B	2.136	5.87	0.716	–9.43	13.7
		Brand C	–1.796	6.654	0.787	–14.91	11.31
		Brand D	3.905	5.859	0.506	–7.64	15.45
		Brand F	–10.056	8.632	0.245	–27.06	6.95
	Brand F	Brand A	26.013*	7.551	0.001	11.14	40.89
		Brand B	12.192	7.118	0.088	–1.83	26.21
		Brand C	8.259	7.777	0.289	–7.06	23.58
		Brand D	13.961	7.108	0.051	–0.04	27.96
		Brand E	10.056	8.632	0.245	–6.95	27.06

**The mean difference is significant at the 0.05 level.*

Source: Primary Data Analysis Using SPSS 21.0.

of availability of resources. These brands manage their resources well as compared to others, which provides them advantage over others. Serving options to choose products helps them to attract and retain customers and create a sense of loyalty among them towards the brands.

Table 5.3 represents the significant differences between various parameters taken in the study, that is, price, quality, variety, color, and Fittings, with respect to casual wears being provided by various fashion brands, Brand-A, Brand-B, Brand-C, Brand-D, Brand-E, and Brand-F, following sustainable practices.

5.6 CONCLUSION

The fashion industry has undergone a revolution because of a shift in client fashion preferences. In addition to catering to the interests of their customers, brands strive to provide creative concepts for the benefit of society. Textile fashion trends have progressed in lockstep with human cultural activities and are progressing further. Nevertheless, the fashion industry could be regarded as the industry with the maximal detrimental effect on the environment. Fashion and textile effluent are released into the drainage system, linked to various ponds, and rivers. Both aquatic and human life can be endangered because of this waste. On that account, growing awareness and knowledge among customers led to the acceptance of sustainable production and consumption practices all over the world. After a dramatic shift in client demand for an eco-friendly product, the fashion and clothes industry began to see a surge in sustainable marketing to satisfy and retain environmentally concerned customers.

This study is directed to understand the satisfaction derived by the consumers based on five different parameters and to what extent these parameters influence their perception towards fashion brands following sustainable practices along with their decision to approach these fashion brands for casual wear. It has been found that the parameters of price, quality, and variety influence the satisfaction and choice of the customers purchasing casual wear from a fashion brand that follows sustainable practices. It has been derived that different parameters of different brands deliver different things. All brands are consistently high priced due to production charges whereas Brand B, Brand E, and Brand F are better in quality in comparison to others. In variety, Brand A, Brand D, and Brand E provide a variety of range to the customers due to the availability of larger resources than others.

KEYWORDS

- **brands**
- **casual wear**
- **fashion**
- **price**
- **quality**
- **satisfaction**
- **sustainable**
- **variety**

REFERENCES

Beard, N. D., (2008). The branding of ethical fashion and the consumer: A luxury niche or mass-market reality? *Fashion Theory, 12*(4), 447–467. doi: https://doi.org/10.2752/175174108X346931.

Bendell, J., & Kleanthous, A., (2007). *Deeper Luxury.* WWF-UK, Godalming. Retrieved from: http://assets.wwf.org.uk/downloads/luxury_report.pd (accessed on 8 December 2023).

Chabowski, B. R., Mena, J. A., & Gonzalez-Padron, T. L., (2011). The structure of sustainability research in marketing, 1958–2008: A basis for future research opportunities. *Journal of the Academy of Marketing Science, 39*(1), 55–70. doi: https://doi.org/10.1007/s11747-010-0212-7.

Chan, T. Y., & Wong, C. W., (2012). The consumption side of sustainable fashion supply chain: Understanding fashion consumer eco-fashion consumption decision. *Journal of Fashion Marketing and Management: An International Journal.* doi: https://doi.org/10.1108/13612021211222824.

Chang, D. R., Jang, J., Lee, E. Y., Lee, H., & Chang, B., (2017). When a good company transgresses: A study of the influences of CSR, moral decoupling, and ethnocentrism. *Journal of Global Fashion Marketing, 8*(1), 40–53. doi: https://doi.org/10.1080/20932685.2016.1255853.

Dekimpe, M. G., & Hanssens, D. M., (2018). Sustained spending and persistent response: A new look at long-term marketing profitability. In: *Long-Term Impact of Marketing: A Compendium* (pp. 143–186).

Eklington, J., (1998). *Cannibals with Forks: The Triple Bottom Line of the 21st Century.* Stoney Creek, CT: New Society Publishers.

Elkington, J., (1997). *Cannibals with Forks: The Triple Bottom Line of the 21st Century.* Capstone.

Fletcher, K., (2013). *Sustainable Fashion and Textiles: Design Journeys.* Routledge.

Gam, H. J., & Banning, J., (2011). Addressing sustainable apparel design challenges with problem-based learning. *Clothing and Textiles Research Journal, 29*(3), 202–215. doi: https://doi.org/10.1177%2F0887302X11414874.

Howarth, G., & Hadfield, M., (2006). A sustainable product design model. *Materials & Design, 27*(10), 1128–1133. doi: https://doi.org/10.1016/j.matdes.2005.03.016.

Joergens, C., (2006). Ethical fashion: Myth or future trend? *Journal of Fashion Marketing and Management: An International Journal. 10*(3), 360–371. doi: https://doi.org/10.1108/13612020610679321.

Joergens, V., (2008). Binary frequency of very young brown dwarfs at separations smaller than 3 AU. *Astronomy & Astrophysics, 492*(2), 545–555.

Jung, J., Kim, S. J., & Kim, K. H., (2020). Sustainable marketing activities of traditional fashion market and brand loyalty. *Journal of Business Research, 120*, 294–301. doi: https://doi.org/10.1016/j.jbusres.2020.04.019.

Kim, J., & Schellhase, R., (2015). Sustainable marketing in Asia and the world. *Journal of Global Scholars of Marketing Science, 25*(3), 195–197. doi: https://doi.org/10.1080/21639159.2015.1044700.

Kim, J., (2015). Sustainability in social brand communities: Influences on customer equity. *Journal of Global Scholars of Marketing Science, 25*(3), 246–258. doi: https://doi.org/10.1080/21639159.2015.1041783.

Knox, S., & Maklan, S., (2004). Corporate social responsibility: Moving beyond investment towards measuring outcomes. *European Management Journal, 22*(5), 508–516. doi: https://doi.org/10.1016/j.emj.2004.09.009.

Ko, E., Hwang, Y. K., & Kim, E. Y., (2013). Green marketing functions in building corporate image in the retail setting. *Journal of Business Research, 66*(10), 1709–1715. doi: https://doi.org/10.1016/j.jbusres.2012.11.007.

Martínez, P., & Del, B. I. R., (2013). CSR and customer loyalty: The roles of trust, customer identification with the company and satisfaction. *International Journal of Hospitality Management, 35*, 89–99. doi: https://doi.org/10.1016/j.ijhm.2013.05.009.

Mazanov, J., & Woolf, J., (2017). Corporate social responsibility and managing drugs in sport. *Journal of Global Sport Management, 2*(2), 96–110. doi: https://doi.org/10.1080/24704067.2017.1314765.

Mittelstaedt, J. D., Schultz, II, C. J., Kilbourne, W. E., & Peterson, M., (2014). Sustainability as megatrend: Two schools of macro marketing thought. *Journal of Macromarketing, 34*(3), 253–264. doi: 10.1177/0276146713520551.

Mora, E., Rocamora, A., & Volonté, P., (2014). On the issue of sustainability in fashion studies. *International Journal of Fashion Studies, 1*(2): 139–147. doi: http://dx.doi.org/10.1386/infs.1.2.139_1.

Murat, A., & Lochard, C., (2011). *Luxury and Sustainability: The New Alliance*. Editions Eyrolles.

Nidumolu, R., Prahalad, C. K., & Rangaswami, M. R., (2009). Why sustainability is now the key driver of innovation. *Harvard Business Review, 87*(9), 56–64.

Niinimäki, K., & Hassi, L., (2011). Emerging design strategies in sustainable production and consumption of textiles and clothing. *Journal of Cleaner Production, 19*(16), 1876–1883. doi: https://doi.org/10.1016/j.jclepro.2011.04.020.

Pfeffer, J., (2010). Building sustainable organizations: The human factor. *Academy of Management Perspectives, 24*(1), 34–45. doi: https://doi.org/10.5465/amp.24.1.34.

Prates, C., Pedrozo, E., & Silva, T., (2015). Corporate social responsibility: A case study in subsidiaries from Brazil and China. *Journal of Technology Management & Innovation, 10*(3), 131–142. doi: http://dx.doi.org/10.4067/S0718-27242015000300014.

Puspita, H., & Chae, H., (2021). An explorative study and comparison between companies' and customers' perspectives in the sustainable fashion industry. *Journal of Global Fashion Marketing, 12*(2), 133–145. doi: https://doi.org/10.1080/20932685.2020.1853584.

Shen, B., Li, Q., Dong, C., & Perry, P., (2017). Sustainability issues in textile and apparel supply chains. *Sustainability, 9*(9), 1592. doi: https://doi.org/10.3390/su9091592.

Shen, B., Wang, Y., Lo, C. K., & Shum, M., (2012). The impact of ethical fashion on consumer purchase behavior. *Journal of Fashion Marketing and Management: An International Journal. 16*, 234–245. doi: https://doi.org/10.1108/13612021211222842.

Shen, B., Zheng, J. H., Chow, P. S., & Chow, K. Y., (2014). Perception of fashion sustainability in the online community. *The Journal of the Textile Institute, 105*(9), 971–979. doi: https://doi.org/10.1080/00405000.2013.866334.

Sheth, J. N., Sethia, N. K., & Srinivas, S., (2011). Mindful consumption: A customer-centric approach to sustainability. *Journal of the Academy of Marketing Science, 39*(1), 21–39. doi: https://doi.org/10.1007/s11747-010-0216-3.

Sun, Y., & Ko, E., (2016). Influence of sustainable marketing activities on customer equity. *Journal of Global Scholars of Marketing Science, 26*(3), 270–283. doi: https://doi.org/10.1080/21639159.2016.1174537.

United Nations World Commission on Environment and Development, (1987). *Our Common Future.* Oxford: Oxford University Press.

Wang, H., Kim, K. H., Ko, E., & Liu, H., (2016). Relationship between service quality and customer equity in traditional markets. *Journal of Business Research, 69*(9), 3827–3834. doi: https://doi.org/10.1016/j.jbusres.2016.04.007.

Watson, M. Z., & Yan, R. N., (2013). An exploratory study of the decision processes of fast versus slow fashion consumers. *Journal of Fashion Marketing and Management: An International Journal, 17*(2), 141–159. doi: 10.1108/JFMM-02-2011-0045.

CHAPTER 6

Using Sentiment Analysis to Identify Consumers' Emotions in the Hotel Industry

VINOD SHARMA,[1] MANOHAR KAPSE,[1] JEANNE POULOSE,[2] and VAISHALI MAHAJAN[1]

[1]*Symbiosis Center for Management and Human Resource Development, Symbiosis International University, Pune, Maharashtra, India*

[2]*School of Business and Management, CHRIST (Deemed to be University), Delhi NCR, India*

ABSTRACT

This research attempted to present a more comprehensive overview of online user-generated data by extending far beyond quantitative analysis. We gathered a distinctive and substantial database of online user ratings for the hotel industry from numerous websites over a significant amount of time. To gauge the quality of hotel service, we divided customer reviews into two categories using the sentiment analysis technique. The impact of those factors in influencing users' overall evaluation and content creation behavior is then investigated. The findings imply that different aspects of user evaluations have considerably diverse effects on how users evaluate products and what motivates them to create content.

6.1 INTRODUCTION

In today's world, organizations use social media platforms to grow their business and improve their customer service experiences. The success and

Sustainability in Marketing Practice: Strategies for Industry 4.0.
Subrata Chattopadhyay, Udit Chawla, (Eds.)

failure of organizations lie in these reviews posted over social media. Positive reviews can only be generated from satisfied customers, which in turn results in motivating prospective customers to acquire services (Gharzouli et al., 2022). Interestingly, more than 90% of hotel bookings are initiated through online travel agents (Bachtiar et al., 2020). Therefore, firms pursue their customers to provide positive reviews of service encounters to encourage customer bookings.

In fact, these reviews, either positive or negative, are very useful for improving customer service experiences if organizations use them effectively. Various studies have been conducted on the analysis of customer sentiments by using sentiment analysis to uncover the hidden message from their reviews (Bachtiar et al., 2020; Liu et al., 2016). There are several stages in sentiment analysis, i.e., data collection, pre-data processing, allotment of weight to each term, classification of items, and finally evaluation.

We used over 37,000 reviews from a hotel website shared by their customers and developed a machine learning (ML) algorithm in R Software to pinpoint the key problem areas. These sentiments (good or bad) have been analyzed through Sentiment Analysis, which would help the management to take corrective measures to increase the overall satisfaction level and create better e-word of mouth for the hotel.

6.1.1 TOURISM AND HOSPITALITY INDUSTRY IN INDIA

The Tourism and Hospitality industry in India has emerged out as one of the most important growth drivers of Indian service sector (Mehta et al., 2023). Considering the rich Indian culture, variety in ecology, and natural beauty spread across the country makes it more significant for the future perspective. Tourism industry in India is ranked amongst the top 10 in 185 countries in the world which contribute to 6.9% to the GDP of the country in 2019. This industry is projected to reach to US $125 billion mark by 2027 (Tourism and Hospitality Industry in India, 2022). It has become the major contributor in the job market by offering around 39 million employment opportunity in the country.

The data confirms that in recent years, India has become the most digitally advanced countries in the world in terms of usages of digital tools for planning, booking, and experience sharing of a journey (Kapoor & Kapoor, 2021). In May 2019, the arrival through e-tourist visa has increased by

21.70% and it is expected to grow by 47% by 2020 (Tourism and Hospitality Industry in India, 2022). The government also realized that India has potential to become the global tourism hub, so it has taken many initiatives to boot the tourism and hospitality industry. This has attracted many international hotel chains in the country and making it very competitive for the local players (Tourism and Hospitality Industry in India, 2022; Hotel Industry in India, 2018–2023).

6.2 METHODOLOGY

Sentiment analysis, also called opinion mining, is the field of study that analyzes people's opinions, sentiments, evaluations, appraisals, attitudes, and emotions towards entities such as products, services, organizations, individuals, issues, events, topics, and their attributes (Pang & Lee, 2008; Lippmann, 1992). As per the author, Liu (2012) Opinions are central to almost all human activities and are key influencers of our behaviors. The backbone of the Service industries is the opinion of their customers regarding everything related to them. With the advancement of ML tools and techniques, it has become easy to measure the views or feedback of the customers and convert them into sentiments. Sentiment analysis had been used by many researchers and industry professionals in various fields, such as movies, music, hotels, restaurants, e-commerce, retails, etc. (Lu et al., 2013; Rui et al., 2011; Zhang et al., 2010).

Sentiment analysis had three basic approaches ML, lexicon-based, and hybrid approach. For the present study, Lexicon method is applied using R. In the present paper using R Package "sentimentr, lexicon, and syuzhet" lexicon method was applied (Jockers, 2017; Halliday, 2013).

6.3 DATA ANALYSIS

One of the best groups of hotels, in USA, had collected the feedback of their customers of last one year. They want to use this feedback to improve their services and enhance their brand value. A total of 30,172 reviews were recorded and manually were labeled by the staff/manager as good or bad feedback. The data consists of only reviews and the labels of the reviews as good or bad.

In order to remove the manual classification of reviews, subjectivity classification approach was used. Using the R Software and package sentiment, different lexicons were used to classify the sentiment and compared with the manual tagged feedback. As per the Package sentimentr several dictionaries are available, out of these dictionaries eight dictionaries were used to find the sentiment of the feedback. These dictionaries are huliu, jockers, rinker, syuzhet, bing, afinn, and nrc (Hu & Liu, 2004; Jockers & Thalken, n.d.; Loughran & McDonald, 2016; Mohammad & Turney, 2010; Rowe et al., 2011; Shu et al., 2016). The results are tabulated in Table 6.1.

TABLE 6.1 Cross-Tabulation of the Feedback's Classification Using Lexicon Methods

Lexicon Method	Classification	Marked Bad Review	Marked Good Review	Correctly Classified Review	Wrongly Classified Review	Neutral Classified
hash_senti-ment_huliu	Negative	13.59	1.17	80.45	19.25	0.26
	Neutral	0.14	0.12	–	–	–
	Positive	18.08	66.86	–	–	–
hash_senti-ment_jockers	Negative	10.3	0.68	77.79	22.19	0
	Neutral	0	0	–	–	–
	Positive	21.51	67.49	–	–	–
hash_sentiment_jockers_rinker	Negative	11.44	0.944	78.631	21.294	0.0688
	Neutral	0.039	0.0298		–	–
	Positive	20.35	67.191	–	–	–
hash_senti-ment_loughran_mcdonald	Negative	13.966	4.65007	74.939	20.5901	4.46
	Neutral	1.918	2.542	–	–	–
	Positive	15.94	60.973	–	–	–
syuzhet	Negative	6.515	0.4772	74.107	25.6492	0.241
	Neutral	0.145	0.096	–	–	–
	Positive	25.172	67.592	–	–	–
bing	Negative	9.535	0.8053	75.781	20.1773	4.0401
	Neutral	2.9265	1.1136	–	–	–
	Positive	19.372	66.246	–	–	–
afinn	Negative	8.518	0.84	76.738	23.247	0
	Neutral	0	0	–	–	–
	Positive	22.407	68.22	–	–	–
nrc	Negative	6.055	1.319	70.522	24.098	5.378
	Neutral	2.999	2.379	–	–	–
	Positive	22.779	64.467	–	–	–

Source: As per the data analyzed using R.

In the lexicon method, each feedback is converted into sentences, then sentiments of each sentence were calculated. Average Sentiment of sentences is considered as the feedback of that particular feedback, which is classified as positive, negative, and neutral. A cross-tabulation of actual marked feedback and classified feedback using different Lexicon methods are tabulated in Table 6.1. After comparing the different lexicon methods, it is found that the method hash sentiment huliu is better for converting the feedback into sentiments.

In order to explore the positive, negative, and neutral feedback, all the feedback which are classified as per lexicon method huliu were analyzed. All the feedback were converted into corpus and document term matrix. A frequency distribution and bar-charts of the keywords, noun, verb, adverb, etc., were created from the positive and negative classified feedback. The keywords distribution table does not make any sense, since most of the keywords are common in positive, and negatively classified feedback. In order to make explore more the POS tagging was used.

Parts of speech (PoS) tagging is a new field in the area of text analytics. Part of speech tagging (POS tagging or PoS tagging or POST), also called grammatical tagging or word category disambiguation, is the process of marking up a word in a text (corpus) as corresponding to a particular part of speech, before tagging some pre-processing of words is also performed, which includes removal of stop words and stemming (Petrov, Das, & McDonald, 2011; Hu & Liu, 2004.

Using udpipe package in R, PoS were identified separately for all the positive and negative feedback already classified using lexicon methods. The highest occurring Nouns, verb, adverbs, and combination are extracted and graphically represented as bar graphs.

6.3.1 POSITIVELY CLASSIFIED FEEDBACK ANALYSIS

From Figure 6.1, word cloud it is clear that the people are talking about, hotels, rooms, staff, stay, night, location, friendly, breakfast, area place, comfort, services, parking, desk, bathrooms, help, restaurants, pool, and so on. These nouns are indicators of lot of good things about the hotels. In order to identify what noun describes in the feedback's adjectives were used. Using adjectives, we are trying to identify what the customers talks about.

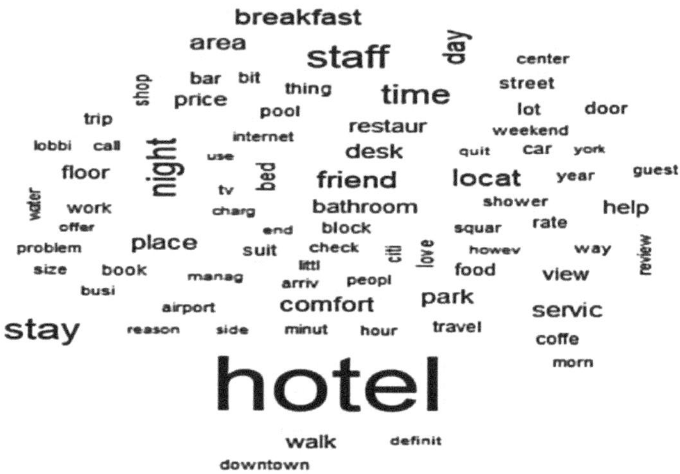

FIGURE 6.1 Word cloud of highest occurring nouns in positively classified feedback.

From the word cloud of adjectives, Figure 6.2 customers are talking great, good, nice, perfect, clean, free, front, small, big, spacious, service, pleasant, modern, extra, enough, and so on. Though these adjectives are not enough to identify the good things about the hotel. In order to find more intuition into the feedback, verb is also used to find the actions. The most common verb associated with positive feedback were stay, get, go, want, bed, need, walk, enjoy, use, felt, eat, visit, and so on (Figure 6.3).

FIGURE 6.2 Word cloud of highest occurring adjectives in positively classified feedback.

FIGURE 6.3 Word cloud of highest occurring verb in positively classified feedback.

Noun and adjectives and noun and verb combinations can give more insight into the positively classified feedback. Bar graph and network charts of the noun and adverbs combination are plotted to understand the feedback.

From Figure 6.4, it is clear that the customers in positive feedback are talking about, flat-screen televisions, DVD players, microwaves, fridge, coffee, and teapots, size of bedrooms and bathrooms, toiletries, bathtubs, and curtains, fresh fruits and breakfast and extremely helpful staff. The combination of noun and adjectives gives lot of insight about the good things of the hotel. Drilling down more into it a combination of noun and verb is also used.

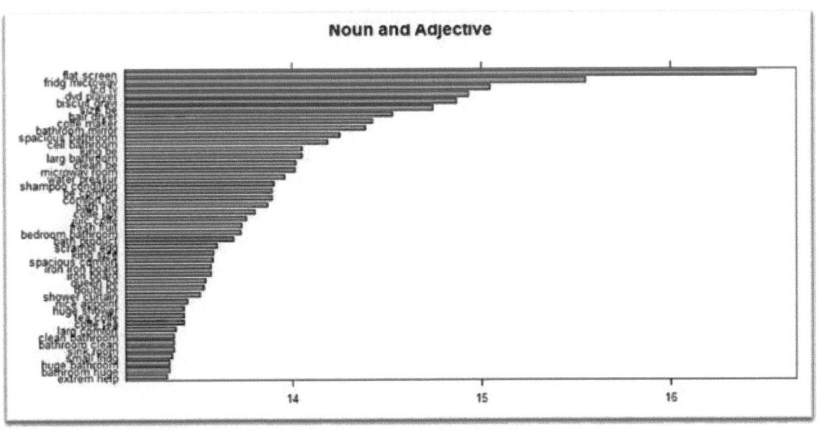

FIGURE 6.4 Bar-graphs of highest occurring combination of noun and adjectives in positively classified feedback.

From the noun and verb combinations graphs from Figure 6.6, it seems that customers are talking about cell phone, credit cards, carry bags, separate shower, hair driers, hold bags, check bags, front staff, glass champagne, coffee makers, etc. (Figures 6.5–6.7).

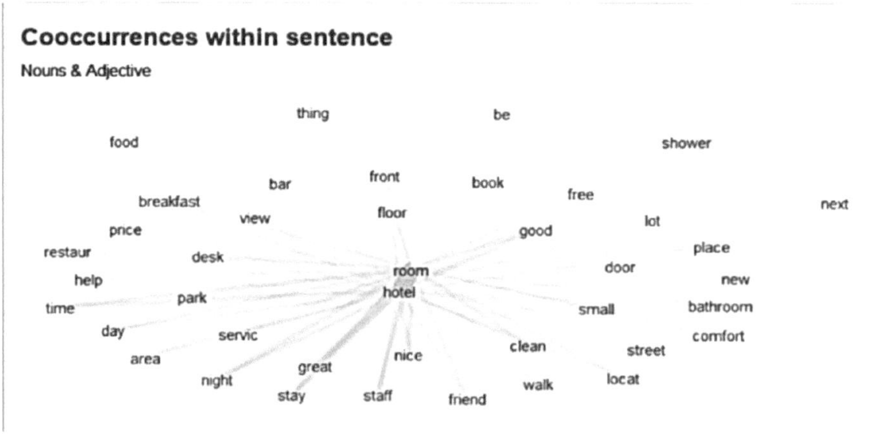

FIGURE 6.5 Network graph of highest occurring combination of noun and adjectives in positively classified feedback.

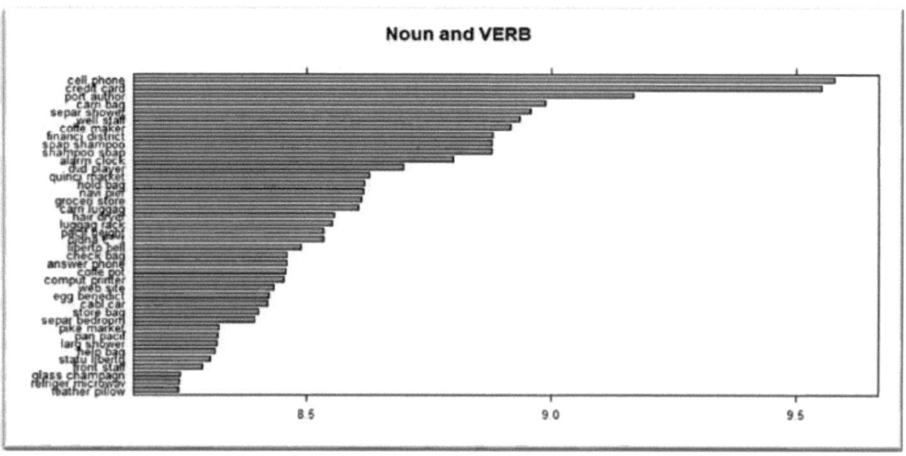

FIGURE 6.6 Bar-graphs of highest occurring combination of noun and verb in positively classified feedback.

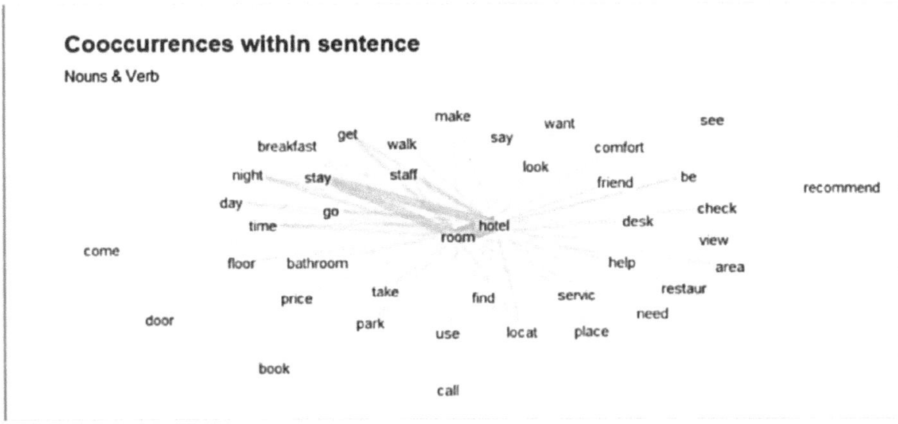

FIGURE 6.7 Network-graph of highest occurring combination of noun and verb in positively classified feedback.

To get more insight into the feedback, phrases are extracted from the same sentences using noun, adjective, and verb combination. From the phrases plotted in Figure 6.8, it is clear that the customers perceptions are positive about front desk staff, flat-screen television sets are liked, location of the hotels are very good, rooms are clean and comfortable, beds, and bedrooms are clean, size of the rooms are comfortable, accessibility of the hotel from other places are good.

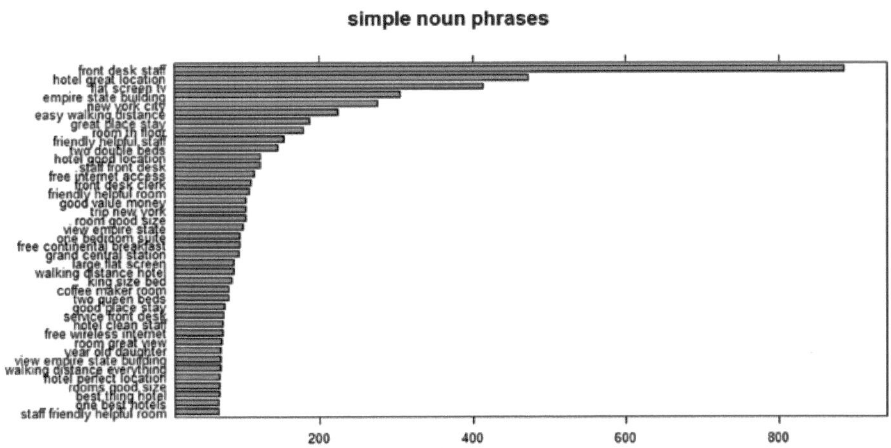

FIGURE 6.8 Bar-graphs of highest occurring simple noun phrases in positively classified feedback.

6.3.2 *NEGATIVELY CLASSIFIED FEEDBACK ANALYSIS*

In order to identify the problems from the feedback, analysis of the negative feedback was done using graphs and charts. Form the most occurring keywords in negatively classified feedback nothing was clear. In order to get a clear insight into the feedback, noun, verb, adverb, and combination of noun and verb, noun, and adjectives were analyzed.

From Figure 6.9, most of the nouns in the negatively classified feedback talks about, room, hotel, night, stay, time, desk, staff, day, place, bathrooms, floor, arrival, money, charge, shower, bed, service, breakfast, hour, water, people, problem, car parking, travel, pool, window, etc. Similarly, a word cloud of adverb is created to understand the problems in a better way.

FIGURE 6.9 Word cloud of highest occurring noun in positively classified feedback.

Figure 6.10 of word cloud of adjectives, it is clear that the customers are talking about front, old, poor, close, smell, loud, key, worst, free, open, dark, services, and so on.

FIGURE 6.10 Word cloud of highest occurring adjective in positively classified feedback.

Though adjectives don't give a lot of insight, a word cloud of verb is created and plotted as Figure 6.11. From the word cloud of verb, it is clear that the customers are talking about stay, check, go, say, seem, said, broken, made, want, pay, and so on. In order to explore more regarding the negative feedback, a combination of noun and adverb; and noun and verb is explored. From Figure 6.12, bar graph and Figure 6.13, network diagram of noun and adjective it is clear that customers are talking about coffee maker, king-size, air conditioners, microwave room, network, internet, bathroom floor and door, towels, cigarette smoke, non-smoker room, valet parking, type of room, room hallway and public transport, these combination of words gives a lot of insight into the problems of the hotel.

FIGURE 6.11 Word cloud of highest occurring verb in positively classified feedback.

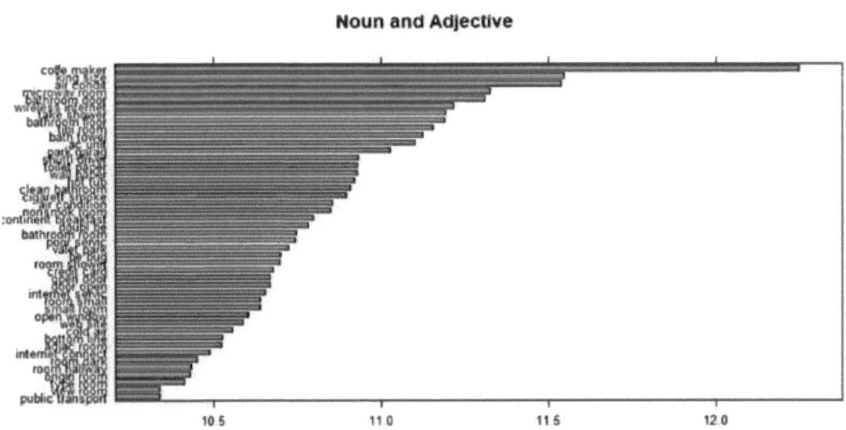

FIGURE 6.12 Bar-graphs of highest occurring combination of noun and adjectives in negatively classified feedback.

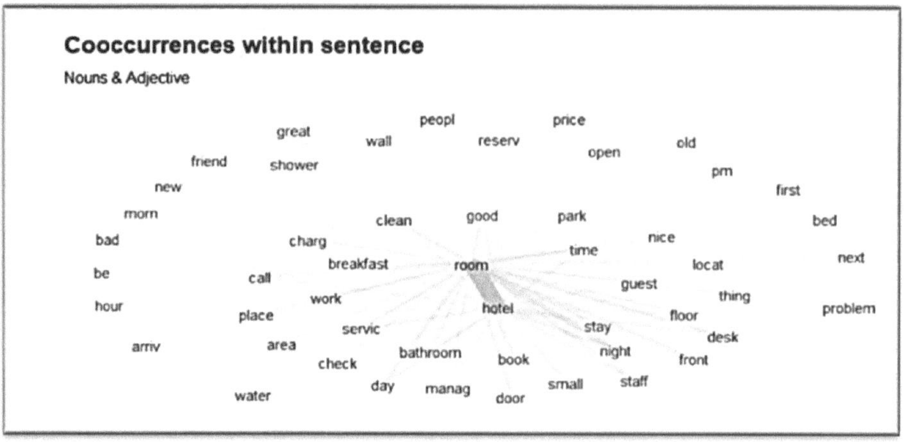

FIGURE 6.13 Network-graphs of highest occurring combination of noun and adjectives in negatively classified feedback.

Similarly, a combination of noun and verb is also studies to get some insight into the customers' feedback so that the problems can be identified. From Figures 6.14 and 6.15 of bar graphs and network diagram of noun and verb, indicates that there is a problem related to credit card, non-smokers rooms, cigarette smoke, valet parking, answering the phone call, parking, stain carpets, bath towels, speaking machines, hair sheet, call machine, taking cab, and so on.

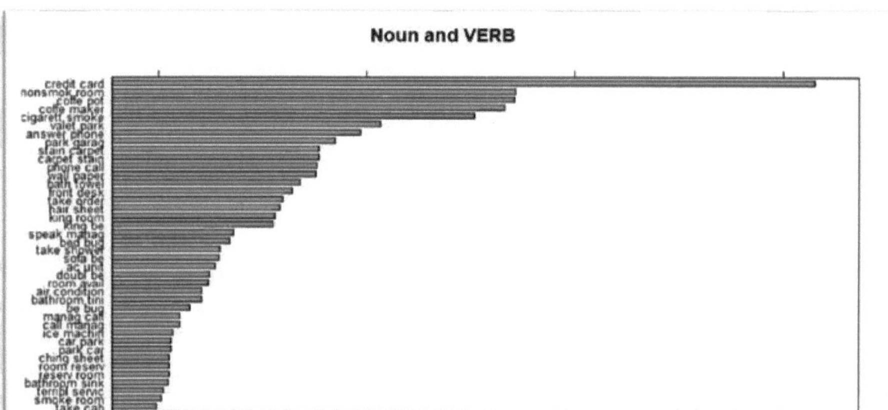

FIGURE 6.14 Bar-graphs of highest occurring combination of noun and verb in negatively classified feedback.

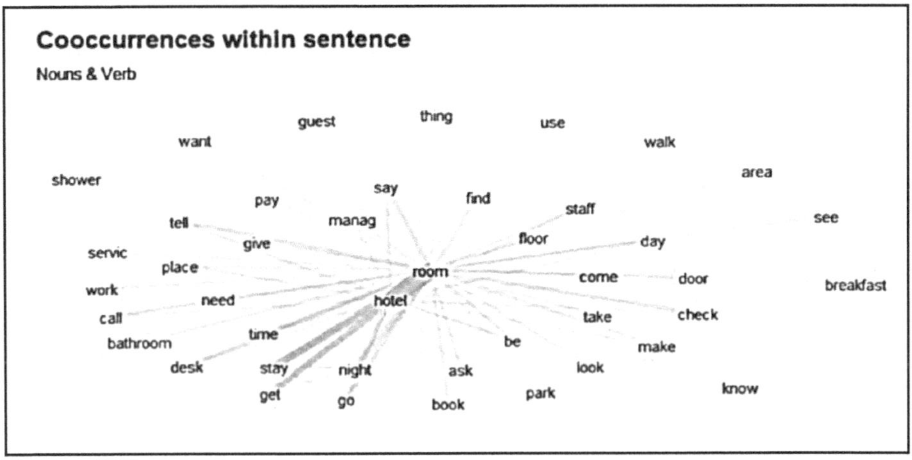

FIGURE 6.15 Network-graphs of highest occurring combination of noun and verb in negatively classified feedback.

Though verb, adjectives, verb, and combination of them gave a lot of information but a simple noun phrase can give more information. For the same a simple noun phrases are created from the same sentences. From Figure 6.16, it is evident that the customers talk about front desk staff, size of the bed, rooms next-door, air-conditioning unit, hotel experience, poor customer service, one-night stay, room service, credit card issues, bedroom bug bite, and front desk complaints.

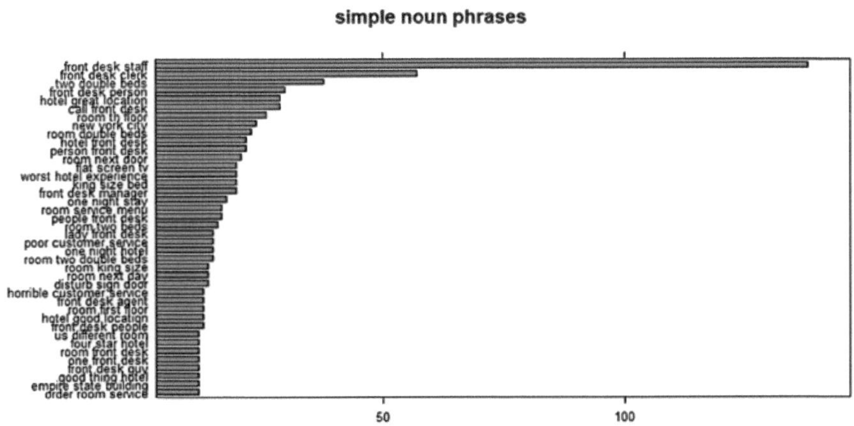

FIGURE 6.16 Bar-graphs of highest occurring simple noun phrases in negatively classified feedback.

From the graphs and charts, it is clear that sentiment analysis and POS can be used to identify the strength and weakness of hotel. These keywords extracted from the feedback can be used to actually understand the problem areas upon which to improve without any human biasedness.

6.4 CONCLUSION

Keywords are the most important tool to understand what is the content of the feedback. In order to understand better this feedback, PoS and the combination of these words are helpful. From the POS tagging, it is clear that there are a lot of problems related to front desk staff employees, double room beds, customer service, credit card, cleanliness in bathrooms, paints, valet parking, shower curtains, public transport, and amenities nearby.

The customers also had a lot of problems with the hallway, cigarette smoke, smoke room, internet connectivity, bathroom towels, adjacent room noise/disturbance, open windows, website, cab facility, and bed bug problem in beds.

Though there are a lot of problems which are identified the hotel also had many good feedback related to front desk staff behavior, location of the hotel, free internet, room size and cleanliness, bed size, bathroom and bathroom size, toiletries, shower, complimentary breakfast items like scrambled egg, crab cake, chicken soup, fresh fruits, etc., size of the television, tea coffee maker, fridge, and microwave in the room, parking, and checking bags.

KEYWORDS

- **consumer emotions**
- **data analysis**
- **feedback analysis**
- **hospitality and tourism sector**
- **hotel industry**
- **sentiment analysis**
- **word cloud**

REFERENCES

Bachtiar, Fitra & Paulina, Wirdhayanti & Rusydi, Alfi Nur. (2020). *Text Mining for Aspect Based Sentiment Analysis on Customer Review: A Case Study in the Hotel Industry.*

Bo Pang & Lillian Lee (2008). "Opinion Mining and Sentiment Analysis," *Foundations and Trends® in Information Retrieval, 2*(1–2), 1–135. http://dx.doi.org/10.1561/1500000011.

Hu, M., & Liu, B. (2004). Mining and Summarizing Customer Reviews. *Proceedings of the Tenth ACM SIGKDD International Conference on Knowledge Discovery and Data Mining,* Seattle, 22–25 August 2004, 168–177.

Hu, Minqing & Liu, Bing (2016). Textual analysis in accounting and finance: A survey. *Journal of Accounting Research, 54*(4), 1187–1230.

Hu, Minqing & Liu, Bing. (2004). Mining Opinion Features in Customer Reviews. *Proceedings of AAAI.*

Jockers, Matthew & Thalken, Rosamond. (2020). Text Analysis with R: For Students of Literature. 10.1007/978-3-030-39643-5.

Kapoor, R., & Kapoor, K. (2021). The transition from traditional to digital marketing: a study of the evolution of e-marketing in the Indian hotel industry. *Worldwide Hospitality and Tourism Themes, 13*(2), 199-–213.

Lei Shu, Bing Liu, Hu Xu, & Annice Kim. (2016). Lifelong-RL: Lifelong Relaxation Labeling for Separating Entities and Aspects in Opinion Targets. In *Proceedings of the 2016 Conference on Empirical Methods in Natural Language Processing*, pp. 225–235, Austin, Texas. Association for Computational Linguistics.

Liu, B., (2012). *Sentiment Analysis and Opinion Mining.* Toronto: Morgan & Claypool Publishers.

Liu, Q., Liu, B., Zhang, Y., Kim, D. S., & Gao, Z. (2016). Improving Opinion Aspect Extraction Using Semantic Similarity and Aspect Associations. *Proceedings of the AAAI Conference on Artificial Intelligence, 30*(1). https://doi.org/10.1609/aaai.v30i1.10373.

Loughran, T., & McDonald, B. (2016). Textual Analysis in Accounting and Finance: A Survey. *Journal of Accounting Research, 54*, 1187–1230. https://doi.org/10.1111/1475-679X.12123.

Lu, Xianghua, Ba, Sulin, Huang, Lihua, & Feng, Yue, (2013). Promotional Marketing or Word-of-Mouth? Evidence from Online Restaurant Reviews (May 30). *Information Systems Research, 24*(3), 596–612. Available at SSRN: https://ssrn.com/abstract=2272131 or http://dx.doi.org/10.2139/ssrn.2272131.

Mehta, M. P., Kumar, G., & Ramkumar, M. (2023). Customer expectations in the hotel industry during the COVID-19 pandemic: A global perspective using sentiment analysis. *Tourism Recreation Research, 48*(1), 110–127.

Mohamed Gharzouli, Aimen Khalil Hamama & Zakaria Khattabi (2022). Topic-based sentiment analysis of hotel reviews. *Current Issues in Tourism, 25*(9), 1368–1375, DOI: 10.1080/13683500.2021.1940107.

Petrov, S., Das, D., & McDonald, R., (2011). *A Universal Part-of-Speech Tagset.* Only arxiv is available arXiv:1104.2086.

ReportLinker. (2023). Hotel Industry in India: Overview, Revenue, Growth Forecast, Trends, Drivers, Challenges, and Data on Tourist Arrival, Distribution Channels, Room Tariff, and Key Players (Report No. 5783881). Retrieved from https://www.reportlinker.com/p05783881/Hotel-Industry-in-India.html.

Rowe, Matthew, Stankovic, Milan, Dadzie, Aba-Sah, & Hardey, Mariann (eds.). (2011). Making Sense of Microposts 2011: Proceedings of the ESWC2011 Workshop on 'Making

Sense of Microposts': Big Things Come in Small Packages: co-located with the 8th Extended Semantic Web Conference, ESWC2011, Heraklion, Crete, May 30, 2011.

Rui, H., Liu, Y., & Whinston, A. B. (2011). Whose and What Chatter Matters? The Impact of Tweets on Movie Sales. *SSRN Electronic Journal*. https://doi.org/10.2139/ssrn.1958068.

Saif Mohammad & Peter Turney. (2010). Emotions Evoked by Common Words and Phrases: Using Mechanical Turk to Create an Emotion Lexicon. In *Proceedings of the NAACL HLT 2010 Workshop on Computational Approaches to Analysis and Generation of Emotion in Text*, pp. 26–34, Los Angeles, CA. Association for Computational Linguistics.

Tourism & Hospitality Industry in India, (2022). https://www.ibef.org/industry/tourism-hospitality-india (accessed on 8 December 2023).

Zhang, Z., Ye, Q., Law, R., & Li, Y. (2010). The impact of e-word-of-mouth on the online popularity of restaurants: A comparison of consumer reviews and editor reviews. *International Journal of Hospitality Management*, *29*(4), 694–700. https://doi.org/10.1016/j.ijhm.2010.02.002 (accessed on 3 January 2023).

CHAPTER 7

An Introspection on FAME India Scheme in the Light of Non-Incentivized Electric Private Four-Wheelers: Potentials and Possibilities

TRIPSHITA SAHA,[1] SHIVAJI BANERJEE,[1] and MOHUA BANERJEE[2]

[1]*St. Xavier's College, Kolkata, West Bengal, India*

[2]*IMI, Kolkata, West Bengal, India*

ABSTRACT

The faster adoption and manufacturing of (hybrid and) electric vehicles (EVs) in India (FAME) Scheme under the Ministry of Heavy Industries as a part of the National Electric Mobility Mission Plan 2020 (NEMMP) is a government initiative to accelerate the process of transitioning to battery-run vehicles as a feasible mass mobility substitute with a green approach for the country. The scheme is segregated into Phase I and II, beginning from 2015 and 2019, respectively, with an initial two-year plan, but they have been extended time and again, also the latter recently extended till 2024. Two Wheelers; Three Wheelers; Public Four Wheelers; Public Buses and Light Commercial Vehicle manufacturers, as well as customers of all these categories, receive incentives from the government at present, the only exception found in the Personal 4-Wheeler category, which is debarred from this scheme. The concerned research effort attempts to point out the needs for providing incentives to Private 4-Wheelers and apprehending the detailed operational structure of FAME Phase I and II to delineate the impact of personal electric car sales on total EV Sales in the country. Moreover, the research identifies particular challenging areas due to the absence of any

Sustainability in Marketing Practice: Strategies for Industry 4.0.
Subrata Chattopadhyay, Udit Chawla, (Eds.)

production or purchase stimuli in the private Electric 4-Wheeler segment and provides certain suggestive actions for adoption in order to reduce the total ownership cost of electric battery-supported vehicles for non-commercial usage.

7.1 INTRODUCTION

"Only advanced battery and registered e-vehicles will be incentivized under the scheme with greater emphasis on providing affordable and environment-friendly public transportation options for the common man."

—Nirmala Sitharaman, Union Budget (2019)

India has been proactive in its efforts towards achieving the sustainable development goals, by responding to various major challenges of the nation; such as issues related to hunger, income inequality, gender inequality, and clean energy. Now, with 10 years to achieve the SDGs by 2030, this decade is considered as the era of action for sustaining the sustainable. One of these actions is reducing carbon footprints by using renewable and rechargeable energy alternatives. Today, the maximum carbon emission in society is caused by existing petrol/diesel-run vehicles on roads, which needs immediate switchover to battery-driven vehicles as a greener mobility substitute (Figure 7.1).

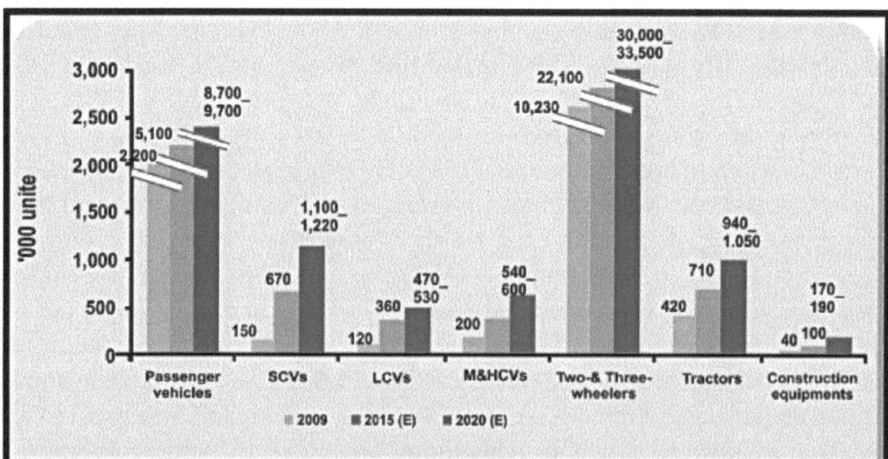

FIGURE 7.1 Projected total vehicle production (in units).

Source: Reprinted from National Electric Mobility Mission Plan, 2020 Report

India took up this ambition of faster EV adoption and manufacturing with the National Electric Mobility Mission Plan 2020 (NEMMP). In conformity with it, The Ministry of Heavy Industries, GOI, National Automotive Board (NAB) designed the Faster Adoption and Manufacturing of (Hybrid &) Electric Vehicles in India (FAME) Scheme. This scheme focuses on providing tax incentives to manufacturers of EVs in five major segments: two-wheelers; three-wheelers (auto); passenger four-wheelers; public buses; and light commercial vehicles. FAME was initially launched on 1st April 2015 with a total fund allocation and expenditure of Rs. 529 crores, Phase-I finally ended on 31st March 2019. Catering to the experience gained from the former cycle and including suggestions from various industry associations and stakeholders, Phase-II of FAME India began on 1st April 2019 with an allocated amount of Rs. 10,000 Cr. over a period of the next three years with incentives on the same vehicle segments as the first phase.

Recently in June 2021, the Government announced to extend FAME-II by two more years, ending it on 31st March 2024 instead of 2022, also a higher amount of incentive on electric 2-Wheelers was declared by 50%. This was purposely done with a motive of allowing a boost in sales of EVs by serving the market deferred demand due to the COVID-19 pandemic scenario. However, the research focus here is the segment of vehicles; i.e., non-commercial or personal electric 4-Wheelers which remains untouched by the government, both under (FAME I and II) in incentivizing and extending tax benefits of any sort to these manufacturers or customers and moreover to analyze the cause and effect of the decision taken on the Indian Automobile Industry at large.

7.2 LITERATURE REVIEW

In the context of the above background, a brief review of literature has been conducted from various available published and gray resources, as per the need of the research:

- **Ministry of Heavy Industries (2019):** This states that "The Phase-I of this Scheme was initially launched for a period of two years, commencing from 1st April 2015, which was subsequently extended from time to time and the last extension was allowed up to 31st March 2019. Market creation through demand incentives was aimed at incentivizing all vehicle segments, i.e., two wheelers, three wheelers (autorickshaw), four wheeler (passenger vehicles), light commercial vehicles, and buses."

- **The Economic Times (2019):** The chapter mentions "The government Friday asserted that subsidies for promotion of electric mobility in India will be only for commercial vehicles, and automobiles for personal usage will not be included. The government has made clear its intention to promote electric mobility through commercial fleet owners and incentives will be only offered to them, Meghwal said here at India UK Electric Mobility Forum 2019."
- Ministry of Heavy Industries, GOI, National Automotive Board (2020): This states that "Government has approved Phase-II of FAME Scheme with an outlay of Rs. 10,000 Crore for a period of three years commencing from 1st April 2019. Out of total budgetary support, about 86% of fund has been allocated for Demand Incentive so as to create demand for EVs in the country. This phase aims to generate demand by way of supporting **7,000 Electric Buses** (e-bus), 5 lakhs electric three wheelers (e-3W), 55,000 electrics four-wheeler passenger cars (including strong hybrid) (e-4W), and 10 lakhs electric two wheelers (e-2W)."
- **KPMG (2020):** This states in their report, "India is the world's fifth largest PVs market, accounting for about 5% of the global PVs sales of 3.4 m units in FY19. However, EV car sales are relatively low. Limited number of models, a lack of charging infrastructure and high-cost differentials are the major adoption challenges. EVs with subsidy are expected to reach TCO parity with petrol cars by 2025. However, if no subsidy is offered, the parity point may be delayed till 2030."

7.3 OBJECTIVES

This study will aim at:

- Primarily, to reckon the influential factors behind a persisting gap in sales of personal electric 4-Wheelers on total EV sales in India.
- Furthermore, understanding the operational structure of FAME – Phase I and II, with the government tax incentives and subsidies provided to the EV vehicle manufacturers.

- Lastly, reflecting the importance and requirement of including personal electric 4-Wheelers in EV tax incentive category under FAME.

7.4 MATERIALS AND METHODS

The concerned research effort is descriptive and symbolic in nature supported with elaborate information gathered from secondary sources as per the need and availability of data. A systematic review of literature based on relevant bibliographic references, such as Industry Annual Reports and gray literature of previously published documentary evidence has been conducted for achieving the above-mentioned objectives. Government Orders, Circulars, Notifications, Operational Guidelines as published by the competent authorities on their official websites and Union Budget Speech 2019–2020, 2020–2021, and 2021–2022 has been referred accordingly, for the same. Qualitative content-based resources from electronic searches and review articles, journals from websites have been utilized considering the novelty of research.

7.4.1 AN OVERVIEW ON FAME INDIA SCHEME

Table 7.1 gives us a brief idea of both the phases of FAME. It also helps us to identify the key varying points between the phases in terms of concerned areas, operational process and targeted goals. On the brighter side the government is seen to be focusing on building technical infrastructure, i.e., charging stations; required to support this mass mobility mode shift. What needs to be observed specifically here is that, though the allotted fund amount has been increased by almost 50% in phase-II, yet the category of EVs to be subsidized is kept unchanged as phase-I, with zero attention on the non-commercial/private segment of EVs.

7.4.2 GOVERNMENT EV ADOPTION INITIATIVES

Figure 7.2 and Table 7.2 both are a summarized view of the incentives provided and subsidies granted by the central as well as few prominent state governments on different segments of EVs, accordingly suited for both commercial and private purpose to the EV manufacturers and users.

TABLE 7.1 FAME (Phase I vs. Phase II)

Specifications	Fame I	Fame II
Commenced from	1st April 2015	1st April 2019
Ended on	31st March 2019	31st March 2024
Objective	To fasten the process of electric vehicles adopting and manufacturing in India.	Emphasized on fastening the process of adopting and manufacturing electric passenger vehicles or vehicles that are mostly used for shared transport, through higher subsidy grants.
Electric vehicles covered	• For two wheelers. • For three wheelers (auto). • For passenger four wheelers. • Buses. • Light commercial vehicles.	• For two wheelers. • For three heelers (auto). • For passenger four wheelers. • Buses. • Light commercial vehicles.
Amount allocated	Rs. 529 crores	Rs. 10,000 crores
Fund utilized	Rs. 529 crores	Rs. 500 crores (till June 2021)
Focused areas	• Demand creation. • Technology platform upgradation (increasing cost effectiveness). • Taking up pilot projects in various cities. • Setting up charging infrastructure (charging stations at frequent gaps in metropolitans and at highways).	• Benefits of incentives to be provided only on EVs supporting lithium-ion batteries. • Incentivizing e-3W and e-4W, registered only for commercial purposes or as public transport. • Incentivizing e-2W registered for both commercial and private usage. • 2,700 charging stations to be established, to find minimum 1 charging station within the range of [3 Km*3 Km] across metros, city connectors and highways.

But personal EV category remains excluded from being incentivized, which puts pressure on the targeted goal achievement for FAME-Phase II, leaving a huge gap in the total EV sales in a particular time frame, as battery driven private cars have enough room to be building a potential market opportunity for the Indian Automobile Industry.

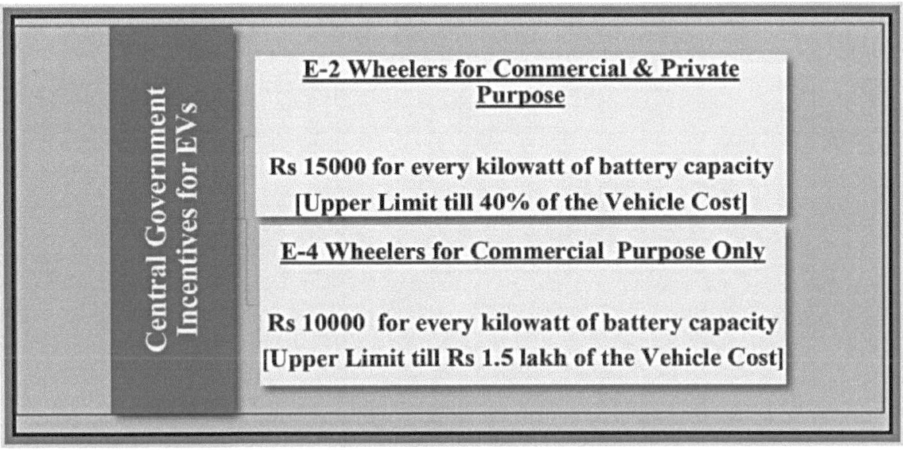

FIGURE 7.2 Central government offerings on electric vehicles.

7.4.3 PRIVATE ELECTRIC 4-WHEELER MARKET IN INDIA

The total electric four wheeler sales in India for past two consecutive financial years, i.e. 2019 and 2020 are 3600 and 3400 respectively. Moreover, this sales figure speaks of an extremely low market penetration rate when compared with the base category of sales. The specific segment market penetration rate is 0.11% in 2019 and 0.12% in 2020 which is (>1) for both the financial periods. However the above information is inclusive of the passenger EV 4 Wheelers for commercial usage (such as cabs, taxis and car rental services) as well as strong hybrid EV 4 Wheelers mostly suited for private purpose, taken together, which does not provide us a very clear picture exclusively regarding sales of private electric 4Wheelers.The KPMG Indian Analysis 2020 mentions, India is the 5th largest market for private vehicles across the world registering a total sales of 34 lakh units in the year 2019 alone. With such a promising existent market opportunity, it can be well apprehended that, there is a huge scope for private EVs too, if effective measures are adopted to bring the total ownership cost at par with regular fuel-driven vehicles, as there is a 2- 2.5 times increased upfront cost difference for owning a private electric four wheeler. Here comes the need for immediate subsidised government policies, innovative EV adoptive models concerning both demand and supply side incentives and government tie-ups with global vehicle battery manufacturers. These outliners are just tip of the iceberg to the plethora of measures that should be initiated by the government to reach a minimum positive (expected 1 to 3%) market penetration

TABLE 7.2 State Government Offerings on Electric Vehicles

State	E-2 Wheelers	E-3 Wheelers	E-4 Wheelers	Other Exemptions
Delhi	Rs. 5,000 for every kilowatt of battery capacity (up to Rs. 30,000) + Scrap value benefits till Rs. 5,000.	Benefits up to Rs. 30,000	Rs 10,000 for every kilowatt of battery capacity (upper limit is Rs. 1.5 lakh).	• No registration fee on EVs. • No road tax on EVs.
Gujrat	Rs. 10,000 per kWh of battery capacity (maximum up to Rs. 20,000).	Benefits up to Rs. 50,000.	Rs 10,000 for every kilowatt of battery capacity (upper limit is Rs. 1.5 lakh).	• No registration fee on EVs. • No road tax on all EVs.
Karnataka	No direct subsidy to EV owners as per SG.	No direct subsidy to EV owners as per SG.	No direct subsidy to EV owners as per SG.	• No registration fee on EVs. • No road tax on EVs.
West Bengal	Rs 20,000 for 2 kWh of battery capacity.	Rs 50,000 for 5 kWh of battery capacity.	Rs 1,50,000 for 15 kWh of battery capacity.	• Nothing announced.
Maharashtra	Rs 10,000 per kWh of battery capacity + Rs. 15,000 early booking discounts + Rs. 7,000 scrap value benefit + Rs. 12,000 other incentives.	Benefits up to Rs. 30,000	Rs 1.50 lakh total discount + Rs. 25,000 scrap value benefit + Rs. 1.5 lakh early booking discount.	• No registration fee on EVs. • No road tax on EVs.
Rajasthan	Grant amount payable Rs. 5,000 for 2 kWh of battery capacity and Rs. 10,000 for more than 5 kWh of battery capacity.	Grant amount payable Rs. 10,000 for up to 3 kWh of battery capacity and Rs. 10,000 for more than 5 kWh of battery capacity.	No direct subsidy to EV owners as per SG.	• State government to reimburse the SGST amount paid on purchase of EVs.
Meghalaya	Grant amount payable Rs. 10,000 for every kilowatt of battery capacity.	No direct subsidy to EV owners as per SG.	Grant amount payable Rs. 4,000 for every kilowatt of battery capacity.	• No fee on registration for EVs. • No road tax on EVs.
Andhra Pradesh	No direct subsidy to EV owners as per SG.	No direct subsidy to EV owners as per SG.	No direct subsidy to EV owners as per SG.	• No registration fee on EVs. • No road tax on EVs.
Telangana	No direct subsidy to EV owners as per SG.	No direct subsidy to EV owners as per SG.	No direct subsidy to EV owners as per SG.	• No registration fee on EVs. • No road tax on EVs.

rate by 2025, by not deferring the FAME-II goals till 2030 without providing any subsidy in the specific segment.

7.4.4 REPERCUSSIONS OF NON-ENTICEMENT ON ELECTRIC PVS

Aforesaid depicts the current scenario of private EVs in Indian Automobile Industry, with an enfeebled sales performance, due to absence of any government incentives, despite a booming market potential that awaits to be captured by bridging the consumer augmented demands. Below pointed out are few of its major adverse effects (Figure 7.4):

Analysis		Four Wheelers	Two Wheelers	Buses	Three Wheelers	LCVs
Ease of Implementation	CAPABILITIES	Low to Moderate	Moderate to High	Moderate	Moderate	Low
	PRICE PERF. GAP	Moderate to High	Low to Moderate	High	High	High
	INVESTMENT	Significant investments required by OEMs	High investment needed as volumes are high.	Moderate	Moderate	Moderate
	Overall	*Low to Moderate*	*Moderate to High*	*Moderate*	*Moderate*	*Low*

FIGURE 7.4 Analysis of implementation of electric vehicle policy.

Source: Reprinted from National Electric Mobility Mission Plan, 2020 Report

1. **Deferred FAME-Phase II Goals:** The target set for 2024, is most likely to defer till 2030.
2. **Negligible Market Penetration Rate:** Private electric cars as compared to 2 and 3 Wheelers, even with subsidy provided, would rate between (1 to 3%) only, whereas the latter two segments will have already penetrated the market with an approximate rate of (7–10%) and (35–45%) respectively, by 2025.
3. **Sinking Sales Figures with Zero Growth:** The lofty total cost of ownership along with an additional upfront cost for electric private cars is almost twice in comparison with the Internal Combustion Engine driven cars on road, this influences the purchase decision of potential private vehicle buyers negatively. As at present, no efforts are being made to bring around the TCO parity for mass acceptance in Indian domestic market.

4. **Private EV Manufacturers in India are Not Optimistic:** To turn around, it is important now, to prepare the market from both demand and supply end. The price factor is directly contributing towards poor demand and negligible sales of E-4W. As an effect, the handful of personal EV manufacturers existing in Indian market, are also not showing enough enthusiasm, outcome being insufficient supply with very limited model ranges, selective model variants and restricted production volume. Also, these could be reasons that might be holding back new private E-4W manufacturing companies from entering and investing in Indian market.

At present India has 13 independent vehicle manufacturers operating in the market and 22 foreign automobile manufacturing company in a joint venture with Indian manufacturers. But, unfortunately due to the above depicted circumstantial shortcomings, only 3 vehicle manufacturers in India till now have come up with electric model variants of private cars. As mentioned earlier, absence of government intervention has created a dearth in both demand and supply ends. Below stated are details for the countable few variants of private E4 wheelers available in Indian automobile industry.

7.4.5 AMELIORATIONS IN NEED

1. Immediate Subsidy for Private E-4 Wheelers: This measure is the stepping stone to fast track the process of personal E-4W adoption in Indian consumer perception (Figure 7.4).

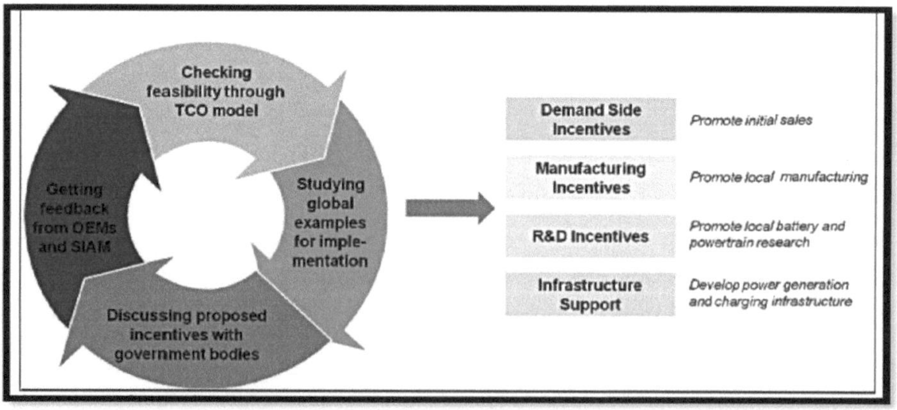

FIGURE 7.4 Total incentives calculation process.

Source: Reprinted from National Electric Mobility Mission Plan, 2020 Report

TABLE 7.3 Personal Usage Electric 4 Wheelers Now in India

E4W Manufacturers	Model Names	Model Variants	Release Date
Mahindra Electric	Everito	[D2 & D6]	June 2016
	E2O Plus City Smart	[P2; P4, & P6]	October 2016
Tata Motors	Tigor EV	[XZ+; XZ+DT & XM]	August 2021
	Nexon EV	[XM; XZ+ & XZ+Lux]	January 2020
Hyundai Electric	Kona Electric	[EE Automatic Premium and EE Automatic Premium Dual Tone]	July 2019 [first electric SUV launched in India]

i. **Demand Side Incentives:** With customer centric focus by reducing the upfront total cost of ownership for bringing a parity with ICE vehicles.

ii. **Supply Side Incentives:** For Inducing manufacturers to invest more for an enhanced production volume and model variations in the private segment of electric four wheelers.

2. **New Model Variants:** Upscale and varied production required of E-4W models in Sedan, SUV, and Hatchbacks with consumer preference and budget friendly options from multiple e-vehicle manufacturers across India.
3. **Infrastructure Goals:** These are to be taken seriously by building more and more charging stations at frequent gaps on highways and metro connectors, roadways. As shown in Figure 7.5, the step-by-step formulation and achievement should be done by considering the types of charger compatibility along the voltage capacity, power (kw) for all the types of EVs, hence has been segregated level wise.

Charging station	Voltage (V)	Power (kW)	Type of Vehicle	Type of compatible charger
Level 1 (AC)	240	<= 3.5 kW	4W, 3W, 2W	Type 1, Bharat AC-001
Level 1 (DC)	>= 48	<= 15 kW	4W, 3W, 2W	Bharat DC-001
Level 2 (AC)	380-400	<= 22 kW	4W, 3W, 2W	Type 1, Type 2, GB/T, Bharat AC-001
Level 3 (AC)	200-1000	22 to 43.5 kW	4W	Type 2
Level 3 (DC)	200-1000	Up to 400 kW	4W	Type 2, CHAdeMO, CCS1, CCS2

FIGURE 7.5 Levels of charging stations and charging compatibility.

Source: Reprinted from NITI Aayog. https://e-amrit.niti.gov.in/standards-and-specifications

4. **Battery Mining Scopes:** Figure 7.6 gives us a clear idea of the Li-ion reserves present in India in comparison with other countries, the portion in terms of global reserves is so much insignificant, that it is not even counted amongst one of the key countries. The scenario is a bit different in case of minerals like manganese, iron ore and titanium, where we find India in the list, but it needs to grow the reserve quantum. The nation needs to start exploring mining opportunities for the key battery components, i.e., lithium and cobalt mainly, for reducing the battery supply dependency in EV manufacturing process and also to develop a significance in global mineral reserves, which can help Indian manufacturers in turn by reducing the upfront total production cost. This will give a scope for creating cost efficiency and enhance reliability in the automobile battery technology.

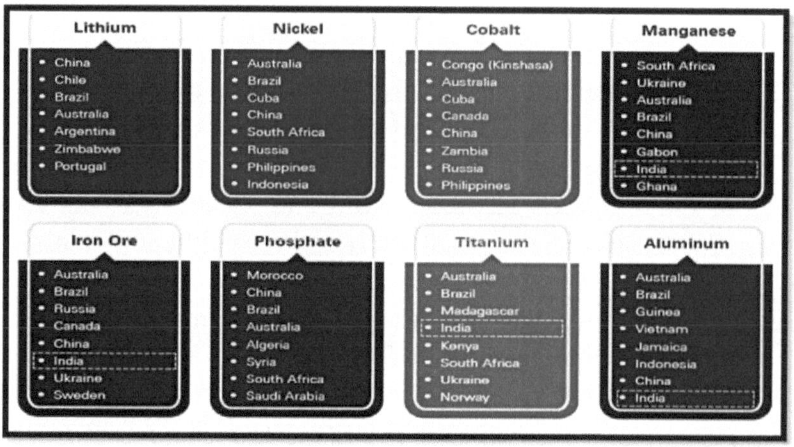

FIGURE 7.6 Mineral reserves in key countries.

Source: https://assets.kpmg.com/content/dam/kpmg/in/pdf/2020/10/electric-vehicle-mobility-ev-adoption.pdf

5. **Battery Swapping Options:** This system to be implemented, requires batteries to be standardized, separately for commercial E-2, 3 and 4 wheelers and personal vehicles, so that the users can exchange the empty used battery with a fully charged battery for a standard rent amount. In order to actualize this, large number of battery swapping stations are to be developed at frequent gaps throughout the country, for reducing any sort of stress to the vehicle owners related to battery draining out probabilities, by this, the users will be able to eliminate the time involved in charging the drained out battery

while commuting, this will also be highly beneficial for commercial vehicle owners, because, battery swapping option will remove the reliability issues with regard to battery quality as all of them will be uniform in variant for specific class of EVs (Figure 7.7).

6. **Battery Leasing and Subscription Facilities:** The main reason for a huge price difference in total upfront cost between ICE and EVs is due to the excessive battery price, which is almost 50% of the total vehicle cost. Now, in order to induce the consumers for purchasing personal E4W, manufacturers can sell vehicles without batteries, and users later can take monthly subscription of batteries or on lease for a fixed time period from the OEMs with a comparatively lower subscription charges according to their usage frequency, urgency, and purpose. This will reduce the vehicle cost to a good extent and will become much affordable for Indian households (Figure 7.7).

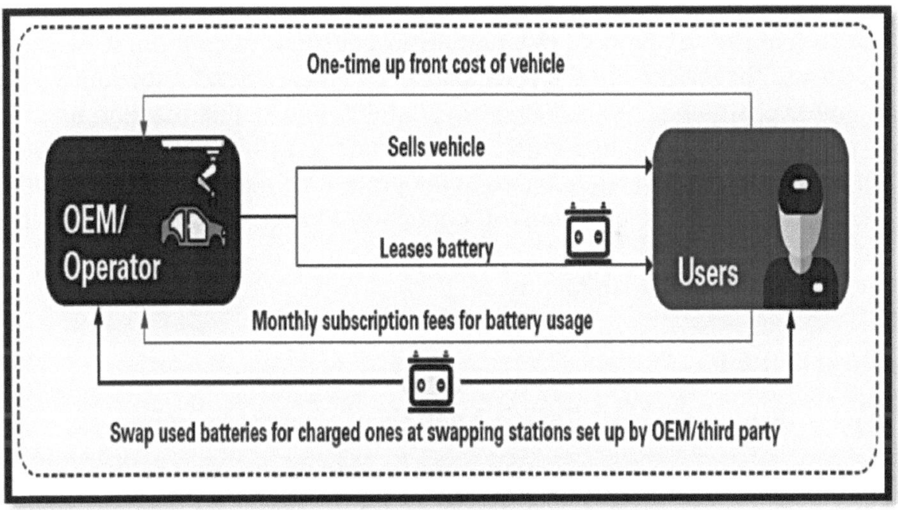

FIGURE 7.7 Battery swapping and subscription operation cycle.

Source: https://assets.kpmg.com/content/dam/kpmg/in/pdf/2020/10/electric-vehicle-mobility-ev-adoption.pdf

7. **Extended Services by OEMs:** The original equipment manufacturers, in order to cater and expedite the private electric 4-wheeler demands will require to provide additional service benefits, than just mere selling of vehicles. This might include vehicle leasing options availability, other than battery leasing, to make consumers accustomed with the product. Apart from that, OEMs can start by creating a charging

community for their customers, where manufacturers would serve their clients with battery charging issues on locality or geographical area basis, at a subsidized energy price, as the government has declared subsidy on electricity, specifically used for EV charging.

7.5 CONCLUSION

The above elaborated are few of the many such action plans, which the government needs to embark on at its earliest to resolve the initial adoption hurdles for private electric four wheelers as compared to ICE vehicles from Indian domestic cerebral. The FAME India Scheme (both Phase I and II) under National Electric Mobility Mission Plan is yet to become one revolutionary initiative as though commercial vehicles and E2, E3 Wheelers, so far are in parallel with the scheme's timeline, yet the private E4W segment despite of a huge market potential, is neither contributing in the goals, nor is helping to expedite the process of EV adoption, rather playing a dawdler, due to inability of product positioning in consumer cognition; absence of incentives for addressing the price barriers; infrastructural uncertainty such as ample charging stations, battery availability and standardization in terms of quality. However, India's EV adoption tale is full of promises, some of which has been delivered, few are in the process of getting addressed, while a significant of them, especially in context of electric private four wheelers are yet to come under the notice of government in an expected recent future for FAME to actualize in totality.

ANNEXURE I

➤ **Segmented Cost Benefit Analysis of Electric Vehicles:**

Analysis		*Four Wheelers*	*Two Wheelers*	*Buses*	*Three Wheelers*	*LCVs*
Total Investments - next 8 years (Rs Crores)	HEV	7,200-7,300	10,000-10,500	1100-1200	500-600	1400-1500
	HEV/BEV	8,700-8,800		1200-1300	800-900	1700-1800
Fuel Benefits	*HG/HEV (fuel saving by 2020)*	*1.1 MT*	*4.9 MT by 2020*	*0.5 MT*	*0.2 MT*	*0.3 MT*
	HG/HEV/BEV (fuel saving by 2020)	*1.6 MT*		*0.6 MT*	*0.3 MT*	*0.5 MT*

Analysis		Four Wheelers	Two Wheelers	Buses	Three Wheelers	LCVs
Ease of Implementation	*CAPABILITIES*	*Low to Moderate*	*Moderate to High*	*Moderate*	*Moderate*	*Low*
	PRICE PERF. GAP	*Moderate to High*	*Low to Moderate*	*High*	*High*	*High*
	INVESTMENT	*Significant investments required by OEMs*	*High as volumes are high.*	*Moderate*	*Moderate*	*Moderate*
Total additional direct new jobs	*Numbers*	*26000-30000*	*22,000*	*1150-1200*	*1200*	*7200-8000*
NPV of benefits (Net Benefits)	**Rs Crores**	**4800-7100**	**28,000**	**3300 - 3700**	**1200-1700**	**1500-3000**

Source: Reprinted from National Electric Mobility Mission Plan, 2020 Report.

KEYWORDS

- **Battery draining**
- **electric vehicle adoption**
- **FAME phase I&II**
- **incentives**
- **NEMMP**
- **personal electric car sales**
- **standardization**
- **total ownership cost**

REFERENCES

Bhardwaj, N., (2021). *Electric Vehicle Industry in India: Why Foreign Investors Should Pay Attention.* https://www.india-briefing.com/news/electric-vehicle-industry-in-india-why-foreign-investors-should-pay-attention-21872.html (accessed on 6 December 2023).

Dalvi, A., (2021). *SMEV: Cumulative EV Sales Down 19.41 Percent in FY 2021.* https://www.autocarindia.com/car-news/smev-cumulative-ev-sales-down-1941-percent-in-fy2021-420595 (accessed on 6 December 2023).

Electric Vehicle Policy, (2021). Power Department, Government of West Bengal, The Kolkata Gazette. (Electric%20Vehicle%20Policy%202021%20(Kolkata%20Gazette%20Notification).pdf (accessed on 6 December 2023).

FAME India Scheme Phase 1 (Faster Adoption and Manufacturing of Hybrid and Electric Vehicles in India), (2019). Ministry of Heavy Industries and Public Enterprises. PTI, https://currentaffairs.anujjindal.in/fame-india-scheme-phase-1/ (accessed on 6 December 2023).

FAME India Scheme Phase II, PIB, (2020). https://www.drishtiias.com/daily-updates/daily-news-analysis/fame-india-scheme-phase-ii (accessed on 6 December 2023).

Garg, A., (2021). *Govt Asserts Subsidy for EVs Only for Commercial Vehicles, Not Personal Usage.* https://www.news18.com/news/auto/decoding-incentives-and-subsidies-on-electric-vehicles-by-various-state-governments-3987410.html (accessed on 6 December 2023).

Government Asserts Subsidy for EVs Only for Commercial Vehicles, Not Personal Usage, (2019). https://economictimes.indiatimes.com/news/economy/policy/govt-asserts-subsidy-for-evs-only-for-commercial-vehicles-not-personal-usage/articleshow/70292418.cms?from=mdr (accessed on 6 December 2023).

McKinsey & Company, (2020). *The Road to 2020 and Beyond: What's Driving the Global Automotive Industry?* McK_The_road_to_2020_and_beyond.pdf (accessed on 6 December 2023).

Ministry of Heavy Industries, (2019). *FAME India Scheme.* PIB Delhi. https://pib.gov.in/PressReleasePage.aspx?PRID=1577880 (accessed on 6 December 2023).

NMEM, National Electric Mobility Mission Plan, (2020). Department of Heavy Industry, Ministry of Heavy Industries & Public Enterprises, Government of India. (nemmp2020.pdf) (accessed on 6 December 2023).

Publication of Notification in Gazette of India (Extraordinary) Regarding Phase-II of FAME India Scheme, (2019). Government of India., Ministry of Heavy Industries & Public Enterprises., Department of Heavy Industry. E__didm_WriteReadData_userfiles_Techanical%20eligibility%20criteria%20under%20FAME-II.pdf (accessed on 6 December 2023).

Shifting Gears: The Evolving Electric Vehicle Landscape in India, (2020). KPMG., electric-vehicle-mobility-ev-adoption.pdf (accessed on 6 December 2023).

Singh, R., & Mishra, S., (2021). *Budget 2021: Push for Domestic Electronic Vehicle Ecosystem Under Atmanirbhar Bharat Needed.* https://indianexpress.com/article/business/budget/electronic-vehicle-industry-india-atmanirabhar-bharat-7167656/ (accessed on 6 December 2023).

Society of Manufacturers of Electric Vehicles in India: EV Market Scenario in India, (2019). https://www.smev.in/ev-sales (accessed on 6 December 2023).

Soman, A., Kaur, H., Jain, H., & Ganesan, K., (2020). *India's Electric Vehicle Transition: Can Electric Mobility Support India's Sustainable Economic Recovery Post COVID-19?* CEEW-India's-EV-Transition-Post-COVID-19-22Dec20.pdf (accessed on 6 December 2023).

Status Quo Analysis of Various Segments of Electric Mobility and Low Carbon Passenger Road Transport in India, (2021). FullReport_Status_quo_analysis_of_various_segments_of_electric_mobility-compressed.pdf (accessed on 6 December 2023).

Tata Motors 75th Integrated Annual Report, (2020). (www.TATA%20MOTORS%20annual-report-2019-2020.pdf (accessed on 6 December 2023).

Taumar, D., & Priya, S., (2020). *Only 10–15% Penetration of Electric Cars is Expected by 2030 in India: Report.* https://auto.economictimes.indiatimes.com/news/passenger-vehicle/cars/only-10-15-penetration-of-electric-cars-is-expected-by-2030-in-india-report/79039710 (accessed on 6 December 2023).

CHAPTER 8

Application and Sustainability of AI and Machine Learning in the Various Stages of the Value Chain in Online Fashion Retail: A Conceptual Review

RADHIKA PILLARISETTY[1] and PRATIKA MISHRA[2]

[1]*Associate Professor, Presidency Business School, Presidency College, Hebbal, Bengaluru, India*

[2]*Professor, Welingkar Institute of Management Development and Research, Mumbai, India*

ABSTRACT

Technology enablers, particularly AI-based tools and processes, have taken the online retail industry by storm. This conceptual chapter outlines the different areas in which machine learning (ML) and deep learning algorithms contribute to enhancing the value chain of online fashion retail processes in such a way that it provides a competitive advantage and increased return on investment. Starting from the product design phase, catalog creation with visual images, manufacturing or retailing, through to customer service with chatbots, and finally the last mile of logistics and delivery of the products, technology empowered by AI is improving the value chain. The contribution of AI techniques such as supervised machine learning, collaborative filtering, k-means clustering, GAN (generative adversarial network), DNNs (deep neural networks), regression models, computer vision (CV) can be directly correlated to increased personalization for customers in giving them products they desire. This directly translates to increased customer satisfaction and

Sustainability in Marketing Practice: Strategies for Industry 4.0.
Subrata Chattopadhyay, Udit Chawla, (Eds.)

loyalty. The chapter brings out the marketing and managerial implications of the use of these intelligent technologies and their sustainability. Increasingly, most retailers, both big and small, are taking advantage of AI-based technologies to improve service satisfaction, sales, and reduce product returns.

8.1 INTRODUCTION

The fashion industry is growing steadily year after year and is valued at over $3 trillion now (Fashionunited.com, 2019). The fashion space today consists of not only clothing, but other accessories like jewelry, bags, watches, sunglasses, etc. Technology enablers particularly AI-based tools and processes have taken the online retail industry by storm. It is allowing the retail industry, especially the fashion industry, to provide an enhanced customer experience and sophisticated tools throughout the value chain. Starting from the inbound processes of product design and procurement to the last mile outbound process of logistics and delivery, AI and different technologies such as machine learning (ML), imaging contribute immensely to the bottom line and increase the ROI of fashion retailers. With technology contributing to the automation of the fashion industry, retailers try to assess the growing needs of consumers in fashion. This will help them to increase their revenue, increase profits, and reduce returns in online fashion retailing. The automation of fashion retailing is helping in reducing the production cycle and providing a wider range of fashion merchandise.

8.2 MACHINE LEARNING (ML)

Machine learning (ML) is an expert system or what we know as artificial intelligence that allows software to give accurate predictions without actually being instructed to do so. This is the AI system learns either from supervised learning or unsupervised learning using training data sets. This training data can be historical data, social media data, or from any other source, but using this data it predicts new output values. ML solves complex problems by trying to imitate intelligent human behavior. There are two types of ML: supervised and unsupervised. Both are used today to solve different problems which were earlier tackled with traditional computational and econometric methods. ML algorithms combined with Big Data today are the base for all searching and retrieval of data which originates from internal sources or external platforms.

8.2.1 STATEMENT OF THE PROBLEM

This study relates to two major domains: Data Science and Marketing. Data Science is an area of computer science that deals with complex algorithms of ML. The scope of this study is to analyze how these algorithms are used in online fashion retail to make the value chain more optimal and efficient. The literature going forward discusses the technical aspects of ML and how it applies to processes in online fashion retail starting from product design to logistics and delivery of products. It tries to bridge the gap between technology and retail. Let us trace the journey of a product assisted by these AI technologies.

8.2.2 PRODUCT DESIGN

During the design phase, it is important to control the cost since the majority of the cost, close to 80%, is determined in the early design phase. We can use intelligent ML systems to help us in our strategic decision making. These algorithms can provide decision support for product engineering decisions and help reduce the risk of new product launches. We can use cluster analysis and scatter plots for accurate predictions to see where the deviations in cost estimates are originating. These models are useful for product managers for the reasons stated above. In the early stages of design, if the product costs are estimated accurately, it helps in arriving at strategic business decisions in the product engineering process (Liebers & Kals, 1997).

One of the challenging aspects of ML problems is that there are many ways to build a model, and one needs computer resources to increase efficiency in model building and choosing the best model. This can no longer be done by humans because of the size of the data. During the design stage as well, ML-based regression analysis can be used to predict the reduction in cost of parts accurately with the help of several qualitative and quantitative models.

We can understand the performance of a 'cross-validated ML framework' using a greedy sequential algorithm for dimensionality reduction. The outcome of this research proves that it is possible to predict the costs of product components in the early design phase with the result yielding an R2 value of 0.960 using only seven features of the product coming from the R&D department. Thus, one can add and remove product features and simulate to observe which features are optimum in terms of product cost and benefits.

8.2.3 SALES FORECASTING AND TREND ANALYSIS

Sales forecasting in retail is very tedious since there are so many products and SKUs across several stores if it is a chain retailer. Nowadays, new algorithms using deep convolutional neural networks (DCNNs), which 'learn' a feature from the time series data automatically and then link the learnt features with weightage given to different features to calculate an accurate forecasting result. This methodology is a combination of several forecasting methods is proving to give retailers competitive advantage over competitors.

8.2.4 CATALOG CREATION AND PRODUCT SELECTION

A subset of ML is a technology called deep learning. This technology known as computer vision (CV) is causing a paradigm shift in online fashion retail. CV can assist customers to choose products through sophisticated deep learning algorithms that provide deep insights into a consumers' buying behavior. Deep learning algorithms are capable of understanding the visual information automatically and these can be very useful in product selection. Fashion businesses today are using technologies such as "Sketch-Product fashion retrieval model" (Jo et al., 2020) which helps the user to draw the product they want. Then using generative adversarial network (GAN) technology it extracts the features of image as vector values and searches for similar products. This personalized recommendation model uses deep neural network (DNN) to calculate the vector values.

8.2.5 COMPUTER VISION (CV) AND IMAGE PROCESSING

ML algorithms using computer vision (CV). CV uses deep learning, a form of ML, where it can 'see' and interpret visual images and in the literal sense gives you 'eyes' on the screen or in a virtual mode where you cannot actually 'see' the product. ML algorithms can 'see' images just like a human brain. Digital signage which is powered by CV can measure customer engagement that will help us in providing better advertising which translates into higher sales revenues. Vision technology can help retailers to analyze the high and low traffic periods in a store and thus having the right promotional campaigns to increase sales.

8.2.6 INVENTORY MANAGEMENT

Today robot powered systems are used for manning and managing the retail store premises, since they improve the overall efficiency inside the retail store and reduce the manpower costs. This is a progression from RFID's which were traditionally used in inventory management. Today technologies powered by AI are being tested and used as in there are mobile robots that can navigate the aisles of the store. These robots are designed to detect shelf out of stock (SOOS) using DCNNs. Predictive analytics helps in keeping the right amount in stock so that your inventory costs are minimized. Inventory robots instead of a physical person can alert the store managers about any low stock or any misplaced item in the store.

8.2.7 PRODUCT REVIEWS AND CRM

Today customer review and ratings have a direct correlation to customers' purchase decision and repeat purchase. A lot of research has been done on the how a positive product review and rating translates to positive perceived product credibility (Li, Huang, Tan, & Wei, 2013). It has been found that customer reviews get better response that expert opinion since product reviews by customers is viewed as helpfulness and helping fellow consumers in decision making. Nowadays, the reviews that are written can also be analyzed using ML algorithms and the result is what is known as Sentiment Analysis.

8.3 SENTIMENT ANALYSIS AND OPINION MINING

Sentiment analysis is widely used on social media comments on companies' websites for product reviews and to gauge the "sentiment" or perception about their products. Text mining is utilized on small snippets of comments from social media (Twitter, Facebook) to check whether or not the text is positive, negative or neutral. This is done using sentiment analysis which analyzes the emotional content of the text. Text analytics algorithms take dumps of raw logs and use certain entities such product, people, topic, event to classify the comments. The context of the text phrases also is important, for example, is the text related to automobiles or paints? There are two basic approaches to sentiment analysis: one is ML, and the other is lexicon-based sentiment analysis (Mitra, 2020). The analysis will yield the results into categories of positive, negative, and neutral comments that can then be utilized for strategic decision making in marketing and advertising campaigns to yield competitive advantage.

8.3.1 VIRTUAL TRY-ONS/VIRTUAL FIT

AI powered digital assistants can recommend apparel and other fashion products to customers based on their size and fit such as height, weight, body type. Customers can get recommendations for styles based on the details they have entered into these size/fit apps. Nike has a similar one for recommending the correct shoe size which is called "NIKE FIT."

8.3.2 CHATBOTS AND CUSTOMER SERVICE

Chatbots that use natural language processing (NLP) are powered by AI and have become very popular and useful. Chatbots have become the drivers of customer service in the pandemic and will continue to provide services such as identifying and qualifying leads 24×7. The purpose of chat bots is to be available anytime to answer customer queries and improve customer experience along the customer journey. It has become a very important marketing tool with increased personalization in local languages to make the customers feel more comfortable in their shopping experience. By collecting the data from the conversations with the customers, chatbots can try and understand customer's personality traits using AI tools. This can improve customer engagement and product recommendations that translate into increased sales and reduced returns.

8.3.3 MERCHANDISING AND SALES

Customer behavior prediction is the key to assessing the market trends and demand for a product. The mathematical tools are only useful when marketers can sense the pulse of the shoppers. Companies like Apple and google can anticipate the needs of the buyers using sophisticated technologies such as heat maps. The cameras in the store along with CV technology will reveal to marketers how their shoppers are behaving inside the store and can study their preferences. Based on this data they can assess the customer engagement and its impact on sales. AI helps in anticipating the buying behavior of the customers and prospects using ML algorithms based on their past purchases, likes, reviews and comments.

8.3.4 DISCUSSION OF RESEARCH FINDINGS AND MARKETING IMPLICATIONS

The technology of CV can assist customers to choose products through sophisticated ML algorithms that provide deep insights into a consumers' buying behavior according to their taste. Using GAN, we have seen that we can generate a combination of fashion (dresses and tops, etc.) in its context of neck design, color, shape, and fit). However, the challenge is that Further, the more deeply we understand consumer behavior we can accurately predict demand and forecast sales. This will help marketers to create products with best features. The new ML algorithms (using DCNN) as we have seen will "learn" the features that can be put in and predict the cost of the product.

Digital signage (powered by CV) can give us data on the type of customers shopping in the store which will help marketers to make better decisions about product promotions and market segmentation. This can then translate into providing an overall shopping experience which will wow the customers. DCNN can also give us accurate sales forecasts that will give marketers an edge over their competition.

Inventory management is another area of focus for retailers to achieve competitive advantage. AI can increase operational efficiency in different areas such as supply chain, inventory management and long-term customer relationship management. Companies like Stylumia, Edited, etc., provide tools and have now been adopted by brands and retailers in India and abroad. Survey of the engineers of these companies has revealed that Stylumia has several high-flying customers that include some of the top brands in fashion.

8.4 PERSONALIZATION OF CUSTOMER EXPERIENCES

Whether it is a small store or a multi-brand retailer, everyone wants to create a personalized experience for the shoppers that is both convenient and enjoyable. With the advent of the pandemic, convenience has taken precedence for customers in that they want to be able to quickly find what they want no matter which mode of shopping they do. That is why today, omnichannel retailing is becoming very popular. During the pandemic, it has become increasingly important to be able to check out items without actually picking up the items for the fear of germs, and technology such as CV can help you see what is in the shopping cart without actually touching it. An intelligent POS system captures data about your purchases, which provides ideas to research for new products and feature enhancement in existing products. Sentiment analysis can churn out information as to what people are discussing about a particular brand, which leads to better decisions in product promotions. Intelligent technologies today at every step can help automate different areas of running a retail store, leaving the store personnel free to provide superior customer service to their clients.

In this chapter, we justify our argument that automation of fashion using AI and ML is being increasingly used and is going to be sustainable since with experience the models will become more optimized and efficient. These tools increase the "touch and feel" quotient in online fashion retail, which translates to lower returns and hence increased customer satisfaction. AI helps all along the value chain with improved predictions and forecasts, better pricing using dynamic pricing, and accurate product placement. This helps customers to have a better shopping experience. Ultimately, it is all about creating a positive perception about the retail brand site where the customer is shopping that creates a competitive advantage.

8.5 CONCLUSION AND FUTURE SCOPE

Today, most retailers, big and small, have adopted these technologies in order to gain a competitive advantage over their rivals and boost sales. Also, social media can harvest the comments that can now be processed instantaneously by these ML algorithms, thus reducing waste and increasing customer satisfaction.

AI can be used to automate CRM activities, predictions about customer buying behavior, and patterns. Today, AI in retail enables the processing of large volumes of data using analytics and ML algorithms. They provide insights into future demand patterns, online product recommendations, etc. The sustainability of this technology lies in the fact that today considerable advances have been made in terms of storing and accessing data, such as Big Data implementations and cloud computing. All this technology helps us to get accurate results across different areas of the value chain using these algorithms. The only challenge is to provide the right kind of training data for these algorithms to work on. However, it seems like AI in retail is here to stay and will only become more pervasive as time goes by.

KEYWORDS

- **chatbots**
- **customer service**
- **deep learning**
- **fashion retailers**
- **generative adversarial network**
- **machine learning**
- **opinion mining**
- **sentiment analysis**

REFERENCES

Agarwal, N., & Yang, S. J., (2014). Social computing, behavioral-cultural modeling, and prediction. *SBP: Lecture Notes in Computer Science, 8393*. Springer, Cham. https://doi.org/10.1007/978-3-319-05579-4_45.

Asiedu, Y., & Gu, P., (1998). Product life cycle cost analysis: State of the art review. *International Journal of Production Research, 36*(4), 883–908.

Eberhart, R., & Shi, Y., (2007). *Computational Intelligence*. Elsevier Inc., Burlington, MA, USA.

Economic Times, (2017). *Fashion Intelligence Startup Stylumia Enters UK & Europe–* ET Retail, [online] https://retail.economictimes.indiatimes.com/news/industry/fashionintelligence-Startup-stylumia-enters-uk-europe/60890605 (accessed on 6 December 2023).

Fashionunited.com (2019). *Global Fashion Industry Statistics* [online] https://fashionunited.com/global-fashion-industry-statistics/ (accessed on 6 December 2023).

Gan, Q., Ferns, B. H., Yu, Y., & Jin, L., (2017). A text mining and multidimensional sentiment analysis of online restaurant reviews. *Journal of Quality Assurance in Hospitality and Tourism, 18*(4), 465–492. https://doi.org/10.1080/1528008X.2016.1250243.

Gupta, N., Abhinav, K. R., & Annappa, (2013). Fuzzy sentiment analysis on microblogs for movie revenue prediction. In: *2013 International Conference on Emerging Trends in Communication, Control, Signal Processing and Computing Applications (C2SPCA)* (pp. 1–4). doi: 10.1109/C2SPCA.2013.6749350.

Jo, J., Lee, S., Lee, C., Lee, D., & Lim, H., (2020). Development of fashion product retrieval and recommendations model based on deep learning. *Electronics, 9*, 508. 10.3390/electronics9030508.

Li, M., Huang, L., Tan, C., & Wei, K., (2013). Helpfulness of online product reviews as seen by consumers: Source and content features. *International Journal of Electronic Commerce, 17*(4), 101–136.

Liebers, A., & Kals, H. J. J., (1997). Cost decision support in product design. *CIRP Annals, 46*(1), 107–112.

Liu, Xuemei, Y., & Qin, Y., (2021). A machine learning-based sentiment analysis of online product reviews with a novel term weighting and feature selection approach. *Information Processing & Management, 58*(5).

Loyer, J. L., Henriques, E., Fontul, M., & Wiseall, S., (2016). Comparison of machine learning methods applied to the estimation of manufacturing cost of jet engine components. *International Journal of Production Economics, 178*, 109–119.

Molcho, G., Cristal, A., & Shpitalni, M., (2014). Part cost estimation at early design phase. *CIRP Annals, 63*(1), 153–156.

Patodiya, P. K., & Birla, P., (2017). *Impact of Virtual – Try – On Online Apparel Shopping Decisions, 42*(2), 197–206.

Philander, Kahlil & Zhong, Yunying. (2016). Twitter sentiment analysis: Capturing sentiment from integrated resort tweets. *International Journal of Hospitality Management. 55,* 16–24. 10.1016/j.ijhm.2016.02.001.

Shuai, Y., & Guiyang, X., (2019). Try it on! Contingency effects of virtual fitting rooms. *Journal of Management Information Systems, 36*(3), 789-822. doi: 10.1080/07421222.2019.1628894.

Sundsøy, P., Bjelland, J., Iqbal, A. M., Pentland, A., & De Montjoye, Y. A., (2014). *Big Data- Driven Marketing: How Machine Learning Outperforms Marketers' Gut-Feeling.* In: Kennedy W.G.

PART III

Sustainable Opportunities

CHAPTER 9

An Empirical Study in Analyzing Customer Preferences and Adoption of Electric Vehicles

M. A. JABBAR,[1] UDIT CHAWLA,[2] VARSHA MISHRA,[3] HENA IQBAL,[4] HARSH VIKRAM SINGH,[5] and SARABJOT SINGH6

[1]*Professor and HOD, Department of CSE (AI & ML), Hyderabad, Telangana, India*

[2]*Associate Professor, University of Engineering and Management, Kolkata, West Bengal, India*

[3]*MSc Digital Marketing, Royal Holloway, University of London, UK*

[4]*Adjunct Faculty, Murdoch University, Dubai*

[5]*Techno India University, Kolkata, West Bengal, India*

[6]*Vellore Institute of Technology, Chennai, Tamil Nadu, India*

ABSTRACT

The types of EVs, the expertise hired, the compensations for internal combustion engine cars, the growth of sales over the previous few years, as well as several charging techniques and possible upcoming technologies, were all observed in this chapter. We also discussed the primary research obstacles and untapped potential. Batteries play a significant role in EVs since they control the car's autonomy. According to these structures, we examined several sequence types. We also acquired knowledge that could be useful in the future, such as graphene, which is predicted to be a solution that makes it possible to store larger amounts of electricity and charges

Sustainability in Marketing Practice: Strategies for Industry 4.0.
Subrata Chattopadhyay, Udit Chawla, (Eds.)

in less time. This kind of data might aid the EV, which could increase its range and increase drivers' and users' acceptance of it. The tradition of the quickest and most prevailing charging methods, as well as additional innovative wireless charging technologies, will be encouraged by the expansion of batteries with larger dimensions. Another element that can help with the disposal of electric vehicles (EVs) is the development of special linking that can be utilized wherever. During the next few years, metropolitan cities will play an important role, these vehicles will perform a major part, thus it will be crucial to have flexible charging regulations that can adjust to the demands of the workforce. As a result, new circumstances brought about by new sequences should be taken into account in the future. Also in the automobile sector, EVs can be seen as an advancement of new techniques that can support sustainability by reducing carbon emissions, improving air quality for residents, and creating employment possibilities with beneficial economic effects.

9.1 INTRODUCTION

An automobile that relies on one or more electronic motors for propulsion is known as an electric vehicle (EV). It derives power from a battery, an accumulator system, or energy from external sources (sometimes generated by solar panels or converting fuel to energy using fuel cells or a generator). Individual transportation habits need to change as a result of the rising price of fossil fuels and the negative environmental effects of their emissions. The industry, which has been dominated by internal combustion engines, is increasingly shifting towards electric vehicles (EVs).

In order to replace combustion engines, which consume a lot of fossil fuels and produce a significant amount of air pollutants that have a major impact on public health and the environment, the world is actively working to develop sustainable and more efficient automotive technology (Silva, 2009; Garcia, 2017). The number of EVs worldwide has surged, from 5,000 in 2008 to more than 2 million in 2016, according to a study by the International Energy Agency (IEA). This is due to several factors, such as growing environmental concerns, declining battery prices, and an increase in the availability of charging infrastructure. Experts have predicted that EV adoption will accelerate rapidly over the next 10 years as a result of all these factors, with growth estimates ranging from 27% to 33% by 2030 (Somayaji, 2017).

We cannot envision a world without EVs. As a result, the power grid is under increased pressure to deliver the necessary amount of electricity without disruptions. The network can be effectively monitored to enable this. The system is protected from any failures by the phasor measurement unit (PMU), which measures power and its instantaneous flow. Using the installed PMU and the power distribution buses, the voltage stability issue has been investigated and resolved in the past (Somayaji, 2017). The strategy for increasing the adoption of electric vehicles is based on two key principles: aggregating demands for similar products that can help scale up operations quickly, and the battery exchange model that may help reduce initial costs for electric vehicles and improve the charging experience. Compared to vehicles with internal combustion engines, EVs are approximately four times more efficient at utilizing the energy they receive to overcome the weight of other vehicles on the road (Barton & Schütte, 2017).

The Indian government has recently taken several steps towards the establishment of sustainable and conveniently accessible charging stations for the benefit of EV owners. Additionally, EV users should have easy access to charging stations that do not contribute to traffic (Deb, Tammi, Kalita, & Mahanta., 2018). Trams have been on the streets since the mid-19th century, but their usage is limited. The latter part of the 20th century witnessed a significant demand for syngas/alternative fuel vehicles to mitigate the emissions issues of internal combustion engines, as fuel prices for fossil fuels increased and reduced dependence on countries that did not have an internal source of crude oil, thus having to import foreign crude oil (Ajanovic & Haas, 2019).

9.2 REVIEW OF LITERATURE

When EVs first became common initially, studies into them failed. Introduce again the knowledge, this period with a focus on reusable energy (Kotecha, 2017). As an outcome of a rise in contamination that has reached a dangerous atmospheric level, there has been a worldwide drive for non-hydrocarbon vehicles. The key to sustainability is reducing pollution levels and circumvention (Gupta et al., 2018; Dwivedi et al., 2019, 2020). Most nations are broadly using substitute renewable options of fuel to control their main transport structures (Ayre, 2016). Scholars (Ahman, 2006 in the Japanese context; Helveston et al., 2015) in the United States and China Researchers are also looking into issues including EV pricing, assortment, charging infrastructure, usability, the overall rate of ownership, cost-benefit, and swapping prices that may impact the acceptance of EV.

The biggest issue preventing EVs from actually being commercially viable in India is a lack of suitable charging infrastructure. In contrast to China, which needed to finish 456 K charging channels in a similar time, India had only 650 channels. Time charges are one more reason that causes anxiety. With reference to currently existing expertise, mobile stallion competence ranges from the low-slung '70s to the upper 90's in terms of percentage (Davis, Rowley, & Carroll, 2013).

An individual takes two factors into account when buying a new car, the price and additionally, the vehicle's performance (Potoglou & Kanaroglou, 2007); Rogers (2004) identified five factors that affect market acceptance (PD) of slightly original produce, one of which is a qualified benefit, that is the extent to which the new knowledge is perceived to be superior to the outdated one; (Bockarjova & Steg, 2014) also claimed that the perceived benefits of using, the high initial price, incomplete charging infrastructure, and short cordless life are assumed to be key barriers to EVs acceptance, though low running costs and environmental benefits are seen as essential motivators (Bockarjova & Steg, 2014; Carley, 2013; Lebeau & Mierlo, 2013). As a result, EVs' initial purchase price is crucial when compared to CVs (Carley, Krause, Lane, & Graham, 2013). When choosing an automobile, consumers frequently overlook the low running costs of the automobiles and consider the high upfront cost as the primary barrier (Hagman, Ritzén, Stier, & Susilo, 2016; Turrentine & Kurani, 2007).

9.3 OBJECTIVE OF STUDY

The objective of this study is to identify the factors influencing consumers' preference toward electric vehicles.

9.4 RESEARCH METHODOLOGY

With the use of an organized questionnaire and normal communications with the consumer, 452 samples were collected from the metro cities of Howrah, Kolkata, and UAE. The data was collected from people of different age groups, gender, occupation, and income. Each parameter studied for the survey was calculated using a scale that points from 1–5.

For the study, a convenience sample was used to gather responses. This method involves taking a sample of individuals or responses from a large population. This sample includes people who are approachable and reflects the total population. The convenience sampling method is a good way to gather information from customers in a market or business setting. It makes it easier to understand what they are thinking and feeling. This study was conducted in multiple locations in Kolkata, West Bengal, and UAE from October 2021 to July 2022. Table 9.1 lists the surveyed areas.

TABLE 9.1 Sample Demographics of Survey Respondents

AGE	**%**
18–25	37.4
25–30	25.2
30–45	16.2
35–40	15.9
>40	5.3
GENDER	**%**
Female	54.9
Male	45.1
EDUCATIONAL QUALIFICATION	**%**
Madhyamik	5.1
HS	21
Graduation	41
PG	26.1
Others	6.8
OCCUPATION	**%**
Service	26
Business	26.1
Student	45.1
Others	2.8
INCOME	**%**
<10K	33.6
10K–20K	20.8
20K–30K	19.5
30K–40K	15.2
>40K	10.9

9.5 FINDING AND ANALYSIS

Kaiser-Meyer-Olkin (KMO) measures of selecting capability (which ranges from 0–1), the numbers which are close to 1 are observed to be finer, although 0.5 is measured to be the least necessity. Table 9.2 shows the KMO value is 0.893, which is more than 0.5 and therefore, factor analysis can be processed further. Likewise, Bartlett's Test of Sphericity shows how strongly the variables are related to one another. The significance of the Bartlett's Test of Sphericity may be observed in Table 9.2, where the sig. value is less than 0.05 (0.000). When equal results are taken into account, they offer the minimal requirements that must be met before doing a factor analysis.

TABLE 9.2 KMO and Bartlett's Test

Kaiser-Meyer-Olkin Measure of Sampling Adequacy		0.893
Bartlett's test of sphericity	Approx. chi-square	3064.333
	Df	153
	Sig.	0

Total variance explained Table 9.3, each factor expresses as a proper value, labeled as eigenvalue, underneath the title 'Total' of 'Initial Eigenvalues.' Framework with an eigenvalue of more than one is measured for further study as they only signify actual value.

TABLE 9.3 Total Variance Explained

Component	**Initial Eigenvalues**			**Extraction Sums of Squared Loadings**		
	Total	**Percentage of Variance**	**Cumulative (%)**	**Total**	**Percentage of Variance**	**Cumulative (%)**
1	6.738	37.434	37.434	6.738	37.434	37.434
2	1.514	8.413	45.846	1.514	8.413	45.846
3	1.12	6.221	52.067	1.12	6.221	52.067
4	0.969	5.382	57.449			
5	0.897	4.985	62.433			
6	0.757	4.203	66.636			
7	0.714	3.967	70.603			
8	0.702	3.899	74.502			
9	0.637	3.539	78.042			
10	0.622	3.453	81.495			
11	0.584	3.243	84.738			

TABLE 9.3 *(Continued)*

Component	Initial Eigenvalues			Extraction Sums of Squared Loadings		
	Total	Percentage of Variance	Cumulative (%)	Total	Percentage of Variance	Cumulative (%)
12	0.534	2.968	87.706			
13	0.48	2.665	90.371			
14	0.426	2.366	92.737			
15	0.406	2.257	94.995			
16	0.349	1.941	96.936			
17	0.285	1.585	98.521			
18	0.266	1.479	100			

The observation states that Factor 1 (Incentive) consists for a variation of 6.738 that is 37.434% of the over-all variation (Table 9.3), Similarly, Factor 2 (Technical efficiency) consists a variable of 1.514 that is 8.413% of the overall variation, Factor 3 (Environmental Friendly) consists a variable of 1.120 that is 6.221% of the over-all variation and as a result, the first four factors account for 52.067% of the total (Table 9.4).

TABLE 9.4 Rotated Component Matrix

	Component		
	1	2	3
V2- Electric vehicles produced fewer emissions than conventional vehicles.	.777		
V1- Electric vehicles are environmentally friendly because they have zero emissions	.739		
V6- There are a lot of benefits or advantages to electric vehicles.	.691		
V26- The government offers different types of financial incentives to make electric vehicles more affordable	.604		
V7- Electric vehicles are much quieter than other vehicles.	.598		
V8- The electric vehicle works on an electric principle.	.588		
V5- In oder to minimize air pollution, Electric vehicles can act as a blessing in lowering emissions	.561		
V3- Electric vehicles can contribute to the environment by saving the future generation.	.525		
V25- The manufacturers of electric vehicles have announced several new models that promise a higher infrastructure	.504		
V14- There are several opportunities for power and battery players in electric vehicles.	.485		

TABLE 9.4 *(Continued)*

	Component		
	1	2	3
V22- The government pushed direct incentives to sow the trend of buying EVs in India		.709	
V20- Vehicles run by electricity are easy to recharge when you are out.		.674	
V17- Electric vehicle technology has improved and they now have a much better range.		.652	
V23- Government-provided subsidy on electric vehicles becomes crucial to smoothen the road for EVs in India.		.534	
V15- The range of Electric vehicles on a fully charged battery is high.		.529	
V10- People want to adopt an electric vehicle because of increased air pollution.			.733
V21- The government seems to have left no stone unturned to encourage people into buying environmentally			.717
V12- Electric vehicles produced fewer emissions than conventional vehicles			.638

The null hypothesis indicates that there is no direct connection between consumer preference and the factors, "incentive," "technical efficiency," and "environmental friendly."

$$H1: B1! = 0$$

The alternative hypothesis indicates that there is a connection, positive or negative, between consumer preference and the factors, "incentive," "technical efficiency" and "environmental friendly."

In Table 9.5, R-square (model summary) value is 0.486 which shows 48.6% of the overall variable in the dependent variable, the total Consumer preference can be enlightened by the self-determining variables, "incentive," "technical efficiency" and "environmental friendly."

TABLE 9.5 Model Summary

Model	R	R Square	Adjusted R Square	Std. Error of the Estimate
1	0.697	0.486	0.483	0.798

We can see from the ANOVA table (Table 9.6) that the regression model predicts the dependent variable significantly well and that its value is statistically significant because the p-value is less than 0.05 (that is 0.000). Therefore, H1 is rejected, whereas H0 is accepted (Table 9.7).

TABLE 9.6 ANOVA

Model		Sum of Squares	df	Mean Square	F	Sig.
	Regression	270.174	3	90.058	141.563	0.000[b]
1	Residual	285.641	449	0.636		
	Total	555.815	452			

TABLE 9.7 Coefficientsa

	Unstandardized Coefficients		Standardized Coefficients	t	Sig.
Model	B	Std. Error	Beta		
(Constant)	3.556	0.037		94.899	0
Incentive	0.14	0.038	0.126	3.73	0
Technical efficiency	0.723	0.038	0.652	19.266	0
Environmental Friendly	0.236	0.038	0.213	6.292	0

$$\text{Consumer Preference} = 3.556 + 0.723 * \text{Technical Efficiency} + 0.236 * \text{Environmental Friendly}$$

The impact of the factor is found to be high towards technical efficiency and environmental friendly followed by others with significant influence towards EVs.

9.6 MANAGERIAL IMPLICATIONS

Ecological concerns and power supplies had encouraged governments to implement a slew of incentive schemes aimed at promoting EVs. As a result of the policy, the worldwide cap of 1 million EVs was surpassed in 2015, with a final tally of 1.3 million. Subsidy schemes were seen to be the most significant and efficient of these incentive schemes. Many nations, including the United States, and Germany, plan to stop providing incentives for EVs. The acceptance of EVs is influenced by a variety of important elements, such as incentive programs and other societal considerations. Policymakers may benefit from understanding these elements in order to develop a better policy framework for EVs.

In several nations of worldwide, electric cars (EVs) are rising in popularity. Compared to ICEVs, EVs are proven to be eco-friendlier and more effective. However, there are not enough charging stations, which limits the global acceptance of EVs. The number of communal locations with EV charging stations is increasing as EVs usage increases. However, using the current electric infrastructure, which is operated by only a petroleum production grid, to charge EVs has an impact on the supply framework and cannot be ecologically beneficial. It might be a wonderful approach and a sustainable move for the ecosystem to charge EVs using photovoltaic panels because solar does have a huge ability to produce power using Photovoltaic panels.

9.7 CONCLUSION

We surveyed the literature on the subject and identified many knowledge gaps and potential research opportunities. The research indicates that there is a high barrier to EV acceptance, and this is due in part to the high price of EVs. There is evidence that EV maintenance costs can be comparable to those of regular cars, but the initial costs are usually higher. Furthermore, there are conflicting results on the effectiveness of government incentives to encourage EV adoption. Further research is needed to determine which government incentives are most effective in encouraging the adoption of EVs, and which problems need to be addressed in order to make this transition a success.

According to research, providing the government with assistance in the construction of EV charging stations could help to build a communal infrastructure and reduce range anxiety. Although preliminary research suggests that the availability of charging stations can significantly impact EV adoption, the source of the correlation is unclear. There is scant research on how governments can best approach the supply of infrastructures, such as through social grants, collaborations between the general populace, government subsidies, or legislation. Currently, there is limited research into the best publicly charging stations, and it's unclear what regional requirements might exist. More research is needed to determine how driving habits and the availability of charging stations relate to the demand for EVs. It is also necessary to integrate policies that impact charging infrastructure and prices with developing technical knowledge about how to charge and operate batteries optimally in order to extend battery life.

The limitations of research on EV uptake are partly due to the reliance on survey data. The body of research shows that people's adoption patterns for EVs differ significantly from what they say they would do. We think it would be valuable to do more research on the size of the gap between what consumers want and what businesses are offering. Research on EVs has a lot of information built up over time, and analysts can use that information to make predictions.

KEYWORDS

- **electric charging infrastructure**
- **electric vehicles**
- **emission**
- **e-vehicle**
- **legislation**
- **policymakers**
- **sphericity**
- **sustainability transportation**

REFERENCES

Ahman, M., (2006). Government policy and the development of electric vehicles in Japan. Energy Policy, 34(4), 433–443.

Ajanovic, A., & Haas, R., (2019). Economic and environmental prospects for battery electric- and fuel cell vehicles: A review. Fuel Cells, 19, 515–529.

Ayre, J., (2016). Top-Selling Car in Norway Is an Electric SUV, EVs 5 29% of Sales. Available at: https://cleantechnica.com/2016/06/13/top-selling-car-norway-electric-suv-evs-29-sales/ (accessed on 6 December 2023).

Barton, B., & Schütte, P., (2017). Electric vehicle law and policy: A comparative analysis. Journal of Energy & Natural Resources Law, 35(2), 147–170. https://doi.org/10.1080/02646811.2017.1262087.

Bick, M. J. (1963). Diffusion of Innovations. Everett M. Rogers. Am. Anthropol., 65, 1146–1147. https://doi.org/10.1525/ aa.1963.65.5.02a00230.

Bockarjova, M., & Steg, L., (2014). Can protection motivation theory predict pro-environmental behavior? Explaining the adoption of electric vehicles in the Netherlands. Glob. Environ. Change, 28, 276–288. https://doi.org/10.1016/j.gloenvcha.2014.06.010.

Carley, S., Krause, R. M., Lane, B. W., & Graham, J. D., (2013). Intent to purchase a plug-in electric vehicle: A survey of early impressions in large US cites. Transp. Res. Part D Transp. Environ., 18, 39–45. https://doi.org/10.1016/j.trd.2012.09.007.

Davis, K., Rowley, P., & Carroll, S., (2013). Assessing the viability of electric vehicle technologies for UK fleet operators. Proceedings of the Universities Power Engineering Conference.

Deb, S., Tammi, K., Kalita, K., & Mahanta, P., (2018). Review of recent trends in charging infrastructure planning for electric vehicles. Rev., Energy Environ., 7(6), e306. Wiley Interdiscipl.

Dwivedi, A., Agrawal, D., & Madaan, J., (2019). Sustainable manufacturing evaluation model focusing leather industries in India. Journal of Science and Technology Policy Management, 10(2).

Garcia, R., & Freire, F., (2017). A review of fleet-based life-cycle approaches focusing on energy and environmental impacts of vehicles. Renew. Sustain. Energy Rev., 79, 935–945. https://doi.org/10.1016/J.RSER.2017.05.145.

Gupta, S. K., Gupta, S., & Gayathiri, S., (2018). Pollution prevention is the key to drive sustainability. Management of Environmental Quality: An International Journal.

Hagman, J., Ritzén, S., Stier, J. J., & Susilo, Y., (2016). Total cost of ownership and its potential implications for battery electric vehicle diffusion. Res. Transp. Bus. Manag., 18, 11–17. https://doi.org/10.1016/j. rtbm.2016.01.003.

Helveston, J. P., Liu, Y., Feit, E. M., Fuchs, E., Klampfl, E., & Michalek, J. J., (2015). Will subsidies drive electric vehicle adoption? Measuring consumer preferences in the US and China. Transportation Research Part A: Policy and Practice, 73, 96–112.

Kotecha, A., (2017). How India Plans to Pave the Way for Electric Vehicles. Available at: https://www.livemint.com/Industry/kDnvPAnuizO1ZSWSGlxtTL/How-India-is-paving-the-way-for-anelectric-future.html (accessed on 6 December 2023).

Lebeau, K., Mierlo, J. V., Lebeau, P., Mairesse, O., & Macharis, C., (2013). Consumer attitudes towards battery electric vehicles: A large-scale survey. Int. J. Electr Hybrid. Veh., 5, 28. https://doi.org/10.1504/ IJEHV.2013.053466.

Peters, A., & Dütschke, E., (2014). How do consumers perceive electric vehicles? A comparison of German consumer groups. J. Environ. Policy Plan, 16, 359–377. https://doi.org/10.1080/15239 08X.2013.879037.

Potoglou, D., & Kanaroglou, P. S., (2007). Household demand and willingness to pay for clean vehicles. Transp. Res. Part D Transp. Environ., 12, 264–274. https://doi.org/10.1016/j.trd.2007.

Silva, C., Ross, M., & Farias, T., (2009). Evaluation of energy consumption, emissions, and cost of plug-in hybrid vehicles. Energy Convers. Manag., 50, 1635–1643. https://doi.org/10.1016/j.enconman.2009.03.036.

Somayaji, Y., Mutthu, N. K., Rajan, H., Ampolu, S., & Manickam, N., (2017). Challenges of electric vehicles from lab to road. In: 2017 IEEE Transportation Electrification Conference (ITEC-India). https://ieeexplore.ieee.org/document/8333880/ (accessed on 6 December 2023).

Turrentine, T. S., & Kurani, K. S., (2007). Car buyers and fuel economy? Energy Policy, 35, 1213–1223. https://doi.org/10.1016/j.enpol.2006.03.005.

CHAPTER 10

A Study on the Growth Prospects of Q Commerce and Its Role Play in eCommerce Business in Present-Day India

JOYSRI DATTA and SUCHANDRA BOSE

Assistant Professor, Faculty of Business Administration Department, University of Engineering and Management, Kolkata, West Bengal, India

ABSTRACT

E-commerce primarily deals with faster delivery of goods to its consumers through e-commerce channels. The study called "Analysis of Delivery Issues that Customers Face upon E-commerce Shopping" by Gupta and Singh deals with the intricate relationship between e-commerce and delivery services. It emphasizes the fact that as e-commerce booms, it depends more and more on the efficiency of the delivery services, which can directly be linked with consumer satisfaction. E-commerce is required to bring speed and efficacy to the delivery services, especially when it comes to essential goods that are perishable in nature, like vegetables, fruits, and certain types of medicines. Since 2020, with the hard hit of the COVID-19 pandemic throughout the world, this study also tried to find out whether consumer satisfaction grew with the limited introduction of e-commerce in certain areas of online businesses and concluded that the satisfaction level did increase to some extent. The study is also a combination of quantitative and qualitative research, which also includes a primary study on subjects between the ages of 18 years to 30 years regarding how they accepted e-commerce and if they think e-commerce actually did contribute to better services. However, one of the limitations of

Sustainability in Marketing Practice: Strategies for Industry 4.0.
Subrata Chattopadhyay, Udit Chawla, (Eds.)

the study is that it could not include candidates from a broader demography. The relevance of e-commerce has become more prevalent as the behavior of consumers is showing a trend of shifting more towards online purchases for both essential and non-essential items, rather than buying goods physically, following traditional means, and this point has been specified by "A Game Changer for the Economy" by CII, back in the year 2016, when the pandemic was not even there, but the ease and comfort of doing business online was very much there. In this study, attempts have been made to understand the relevance of e-commerce in today's society and whether it has contributed towards the betterment of e-commerce and brought in consumer satisfaction. E-commerce can also introduce sustainable practices when conducting businesses online throughout the world by optimizing delivery time. It can also help to serve the community better, aiding the ecosystem especially in businesses for restaurant and food delivery, also supporting SMEs with tech advancements.

10.1 INTRODUCTION

10.1.1 NEED OF Q COMMERCE

Q Commerce is nothing but a model that helps to make a business better by quickening the delivery. Q Commerce is specifically applicable for goods that need faster delivery due to their perishable nature or very limited shelf life. However, if the stereotype is to be broken, Q Commerce can be applied to even typical grooming products like cosmetics, beauty items, hygiene products, pharmaceuticals, etc. (that happened in the UK).

In the COVID scenario, the booming need of Q Commerce can be perceived from quite a distance, and if spoken from the Indian context, few companies have already started implementing the model with baby steps. Goods that are essentially categorized under 'necessities' can now be delivered by eCommerce companies by adopting the Q Commerce model. Thus, eCommerce and Q Commerce are moving hand in hand, depending on the efficacy of one another. Goods that are to be consumed fast, like vegetables, groceries, etc., and goods that are important for survival like life-saving drugs, etc., can be delivered without much complexity following the eCommerce model.

In this study, attempts are being made to understand the success of implementation of Q Commerce in the Indian context and its probable scope in the future.

10.2 LITERATURE REVIEW

E-commerce has not only been a way of convenience for consumers but soon became a channel of promoting products. A good example can be the limited edition of Money Heist Pepsi Cans. As per the Hindu, the company chose to deliver such cans exclusively through Swiggy Instamart (Quick Commerce). Fact is, Grofers and Swiggy Instamart both reported more than one lakh transactions per day over the Quick Commerce channels. The exact estimated growth of these Quick Commerce Channels can be difficult to predict, but as per the Redseer estimation, it does have a potential of growing 10 to 15 times over the next five years (Verma Ambwani & Narayanan, 2021). In the study called "A Review of Management and Importance of eCommerce Implementation in Service Delivery of Private Express Enterprises of China," the authors have tried to observe similar trends in the name of "express delivery." It is basically a review article that deals in the effectiveness of managerial processes revolving around the quality-of-service delivery and customer retention. This chapter also did a study on the literature that highlights the mediation of automation in Supply Chain Management when it comes to strengthening the eCommerce services, and again, depending on the efficacy of fast delivery model (Farooq et al., 2019).

An article in Science Direct called "The Impact of eCommerce on Final Deliveries: Alternative Parcel Delivery Services in France and Germany" stressed on the success of dedicated delivery services.

People shifting from the conventional way of purchasing goods to online mode actually paved the way for more effective and efficient parcel delivery services, that may indirectly contribute to Q Commerce in the future, if not already. The chapter compared the alternatives (against home deliveries) being brought in by French and German Operators that involved pick up point and stores and automated locker networks (Morganti et al., 2014).

According to the Business Standards, India's Quick Commerce market will be reaching $5 billion by the year 2025, and this report has been published by the consulting firm called RedSeer. As per Business Standards, Quick Commerce revolves around delivery of consumables, that too within a span of just 45 minutes, along with a very nominal delivery charge. However, depending on the eCommerce website and the kind of consumables to be delivered, delivery charges can be kept free by the company to boost the sales volume (that could be observed with Flipkart-selected items even on quick commerce) (Alawadhi, 2021). In this regard, a very interesting article called "Analysis of Delivery Issues that Customer Face upon eCommerce

Shopping" published in the International Journal of Management Studies, discusses about the high dependence of eCommerce on the delivery services and thereby satisfying the consumers. Any wrong step in the delivery process can lead to loss of customers. This particular study deals with responses of the customers and their satisfaction level based on various factors of the services, that this study will also aim to cover for a specified demography (Gupta et al., 2019).

As per "A Game Changer for the Economy" produced by Confederation of the Indian Industry (Knowledge partner Deloitte) from the year 2016, eCommerce showed consistent signs of rapid booming. The retail sector has been showing an impressive trend of 11% CAGR displaying significant growth. In this regard, the Indian tele-density also requires special mention, where India exhibited itself to be the second largest market for smartphones, after China. On top of that, it should also be noted that there is a shift even in the usages of mobiles, from voice to data consumption. eCommerce market if observed from Indian Context has the potential to grow rapidly. With the introduction of improved connectivity in both rural and urban areas, will help to boost this trend furthermore ("E-commerce in India | Deloitte India | Technology, Media and Telecommunications," 2022). The article called "Detecting the Crisis of Supply Chain Management on eCommerce for Sustainability Using Q-Technique" talks about sustainable management of eCommerce.

The role play of this study was to understand the application of "Q Technique" in relation to the supply chain crises when it comes to the imported food industry during the situation of COVID-19. With the help of questionnaire survey, the study showed that Q Technique can detect the crisis of what we can call Cross Border Supply Chain Management determining their strength (Ma et al., 2021). As per "Propelling India towards Global Leadership in eCommerce" (by www.pwc.in) India will soon become a digitally empowered society. NASSCOM (along with PwC) highlights the regulatory as well as infrastructure related challenges that the eCommerce players basically face and put forth recommendations to the Government to overcome these challenges. Needless to say, when so much potential in eCommerce can be observed for the Indian context, delivery related complexities and challenges are bound to arise, that needs to be addressed when exploring this area in future ("Propelling India towards Global Leadership in eCommerce," 2018). The Grocer (UK) also lists down important personalities that are involved in faster delivery services. This proves the future potential of Q Commerce and its relevance. It says Q Commerce is not just the game of speed, but also requires strategies and differentiation to win (Nott, 2021). As per Retail Insight Network, Deliveroo actually made an announcement where products related to hygiene and beauty will be delivered following the principles of Q Commerce with a Pilot Scheme that will cover 14 UK cities in the article "Deliveroo and Boots partnership sets new heights for Q-Commerce" (GlobalData Thematic Research, 2022).

10.3 OBJECTIVES

1. This study attempted to explore the need of Q Commerce in present times and their functionality in the current Indian scenario. Q Commerce has already stepped into the country with the help of some online platforms and companies, but even though it is still not that widespread, it is gaining popularity eventually in the Indian retail market.
2. We have tried to understand if the consumers are satisfied at a better level with the introduction of Q Commerce along with the fact that how effectively is Q Commerce successfully functioning in the market and contributing to value addition at present.
3. With this study, we have attempted to identify the areas (product lines) where Q Commerce is functioning currently and also to find out areas where Q Commerce has a potential to function.

10.4 RESEARCH METHODOLOGY

This research can be called explanatory research with the elements of Quantitative Research. A basic primary study has been conducted with 64 respondents who answered regarding their thoughts and experiences regarding Q Commerce. We tried to understand if the participants in our survey think that Q Commerce actually brought betterment in their experiences by shortening the timespan of delivery services and also, the quality of services Q Commerce was able to render so far. This study also attempts to explore questions regarding the relevance of Q Commerce in India and the way it functions in the Indian market.

10.5 DISCUSSION AND FINDINGS

It can be seen that the evidence of Q Commerce is limited in a country like India, but with the help of gathered responses, it can be safely stated that it does carry a fair amount of potential for future business entities. Q Commerce has been applied in very selected product lines so far but its application can definitely be increased for a range of other products like apparels, consumer electronics, etc., in the near future.

Q Commerce is actually responsible for reframing the Indian Grocery Industry (in reference to FMCG products). It shows its relevance especially when it comes to unplanned purchases. Keeping in mind the traditional way of shopping, shoppers needed to have a plan regarding the time and distance as in when to go and where to go to get hold of the necessary items, but with the arrival of Q Commerce, this wave of thinking is eventually shifting as customers can make purchases for essential items (or even non-essential items) whenever the need is recognized. However, in this regard, it is to be kept in mind if the Q Commerce model is working effectively in the proper way as it should have been to establish its prominence in the Indian retail market in the times to come. Q Commerce immediately brings up the concept of 30–45 minutes' delivery and it is especially relevant for mid to high income group households without even requiring the family members to get out of the house physically which is a big convenience for shoppers. Now, coming to the success factors, particularly in India, Flipkart Quick has been introduced which is currently very much at a functional state. Other than that, start-ups like Zepto, Dunzo, Blinkit, Swiggy, Big Basket, etc., are also getting recognition for bringing in Q Commerce in the Indian scenario. In today's era, the lifestyle of common men has become fast as compared to the previous generation. Therefore, given this fast lifestyle, Q Commerce can be said to be the most relevant services that business organization can offer to the public.

Moreover, the required items can actually be obtained easily from one or two online platforms without bearing the hassles of physically visiting more than one store to purchase the desirable items. As per South Asian Newsletter writer, Nilesh Christopher, he too has fallen for Q Commerce, that he detailed in his blog "Quick commerce is reshaping user behavior in India — and that's a problem" (Christopher, 2022).

The research from RedSeer has already been talked about, and as per that research the Indian Q Commerce is expected to grow manifolds (as many times as 10 to 15) that will soon become a $5 billion industries by the year 2025 (Bhatia, 2022). Along with these predictions about the future in regard to the Q Commerce, it is important to find out whether Q Commerce is being welcomed in India or not. As pointed out earlier, there are many start-ups and online platforms operating in India that already indulged into Q Commerce and started serving the customer base. However, even though it is still questionable whether these organizations will be able to bring its services in the rural India anytime soon, one thing that can clearly be observed is that Q Commerce is being appreciated by a great number of online shoppers.

This chapter attempted to capture and analyze the thought process of the shoppers of 21st century and conducted a brief survey regarding their concept of online shopping requirements and Q Commerce requirements, the gist of which is laid down below. Convenience Sampling method has been used in this survey and number of subjects chosen is 64, majorly between the age group of 20 years to 50 years. Here are some of the important highlights that the participants' responses confirmed.

Participants' age, gender, and occupation are given in Figures 10.1–10.3, respectively.

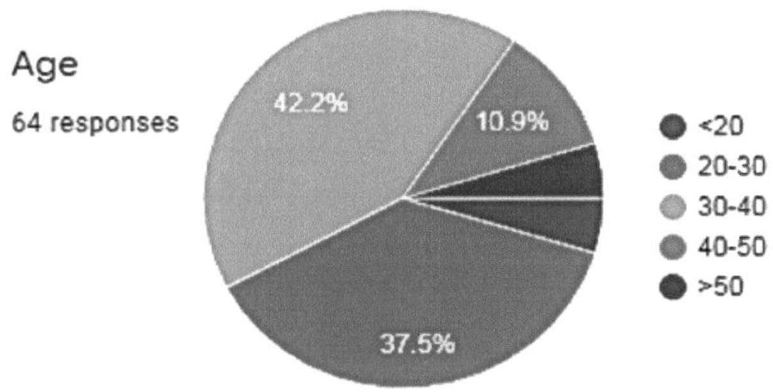

FIGURE 10.1 Age distribution.

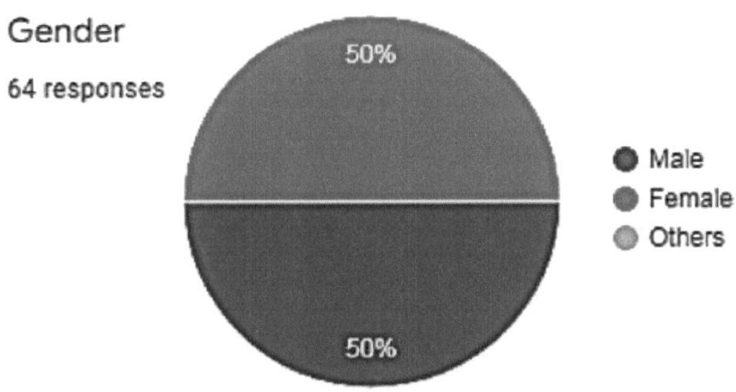

FIGURE 10.2 Gender distribution.

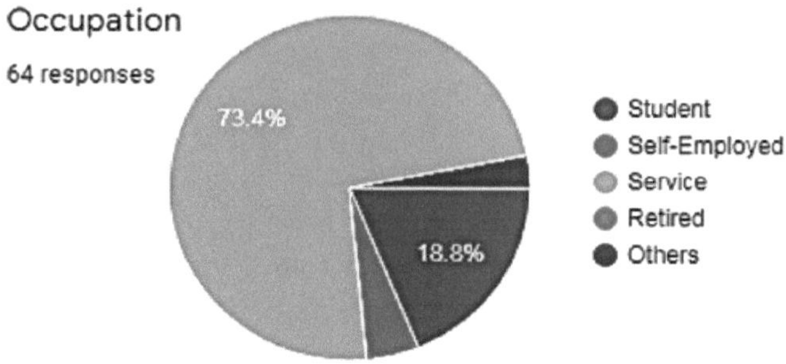

FIGURE 10.3 Occupational distribution.

The study mainly focused on metropolitan area (Kolkata and suburban region) for its subjects to respond as signified in Figures 10.4 and 10.5 (that can also be tagged as 'Tier I' and 'Tier II' cities):

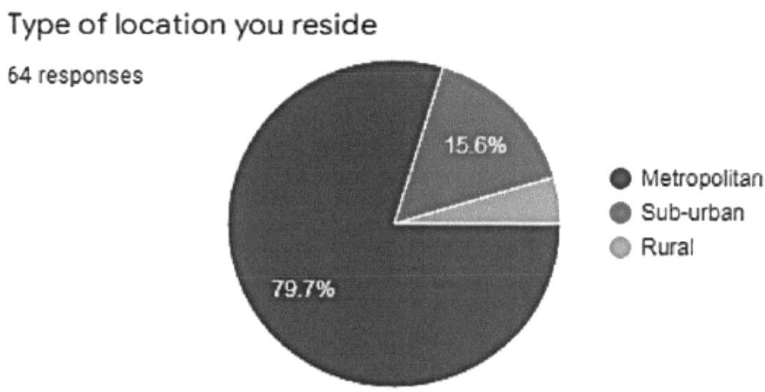

FIGURE 10.4 Area-wise distribution.

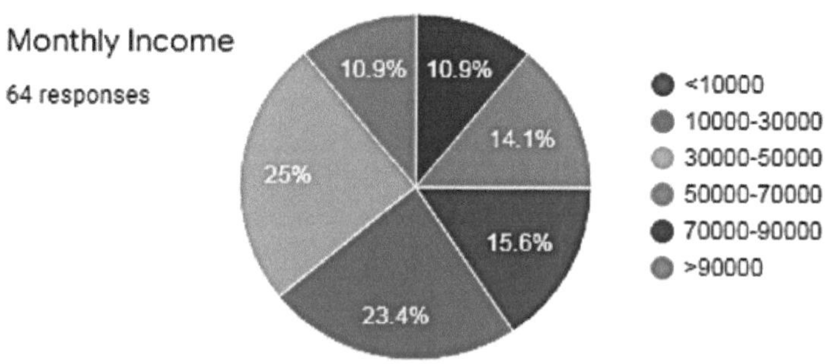

FIGURE 10.5 Monthly income-wise distribution.

Now, some of the highlights from the responses received are shown in Table 10.1.

TABLE 10.1 Results of the Survey Highlighting the Options That Received Maximum Responses Against Each Survey Question

Question Asked	Option 1	Option 2	Option 3	Option 4
You mostly shop from	Online stores (20.3%)	Brick and mortar stores (8%)	Both (71.9%)	–
Do you prefer to buy quick-need items like groceries, daily-need goods, personal care, home care, packaged foods, beverages, over the counter medicines, etc., online?	Yes (53.1%)	No (23.4%)	Maybe (23.4%)	–
How frequently do you buy quick-need items like groceries, daily need products, personal care, home care, packaged foods, beverages, over the counter medicines, etc.?	Monthly (31.3%)	Weekly (21.9%)	Bi weekly (10.9%)	As when need arises (35.9%)
Do you think that shopping from online websites (marketplaces) are a good option for unplanned purchases in case of quick-need items like groceries, daily-need products, personal care, home care, packaged foods, beverages, over the counter medicines, etc., getting delivered within 30 minutes?	Yes (65.6%)	No (5%)	Maybe (29.7%)	–
According to you, do you think that availability of various quick-need items all at one place (online marketplace) provides ease of purchase and saves your efforts?	Yes (81.3%)	No (1.6%)	Maybe (17.2%)	–

The Study also makes attempts to find out the factors that encourages people to go for online shopping rather than to opt for traditional physical shopping. So, the plus points of online shopping are documented as below for the below given question:

"On a scale of 1 to 5, rate the factors which urge you to purchase quick-need items online (1 refers to strongly disagree (SD) and 5 refers to Strongly Agree)" (Figure 10.6).

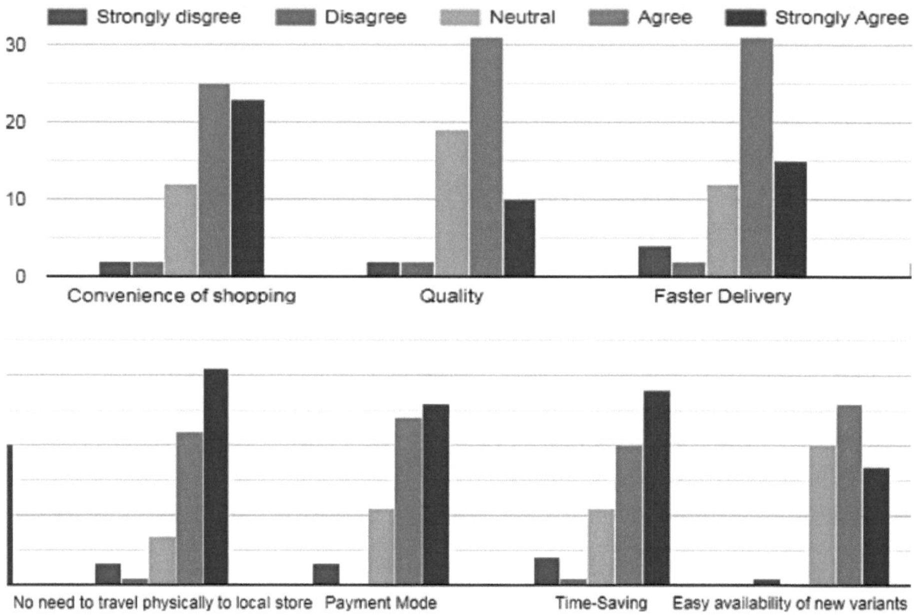

FIGURE 10.6 Factors influencing customers to purchase quick-need items online.

From the above depiction of charts, it is quite clear that there are more than one factors for which the potential customers are opting more and more for online shopping nowadays and Q Commerce surely is acting as a fuel to give eCommerce a perfect boost in this regard. Some of the important factors (as shown above) in regard to why people are going for online shopping is:

- Faster delivery;
- Convenience;
- Quality;
- New variants;
- Ease to pay;
- No physical travel required;
- Time saving.

10.6 CONCLUSION

From the above discussion, it is quite understandable the relevance of Q Commerce in the present society, especially where eCommerce is in full operation. The important points that are required to be highlighted in this case are the follows:

- Most number of our subjects are inclined to shop from both online platforms and Brick and Mortar Stores, however, if only sole online shoppers are taken, it will be more than a whopping 20%, where only Brick and Mortar Stores customers are merely 8%. In this case, it is understandable that the number of online shoppers is fairly more and thus paving the way for Q Commerce (Figure 10.7).

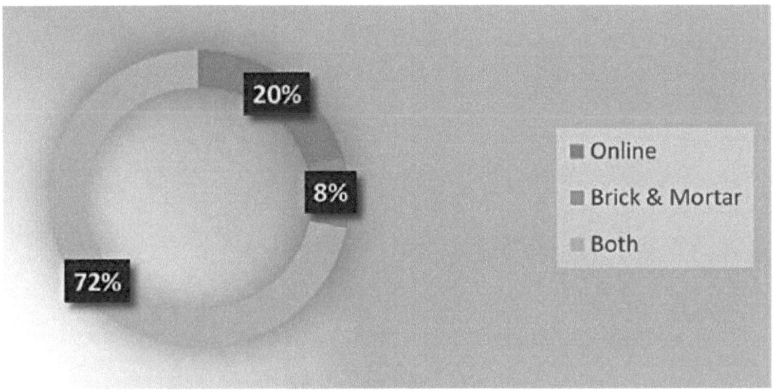

FIGURE 10.7 Preference of shopping mode of customers.

- More than 50% of the consumers say that they prefer online shopping websites for purchasing daily needs and personal care goods, including medicines. Therefore, as far as medicines are concerned, it is a known fact that the involvement and requirement of Q Commerce suddenly becomes high to gain consumer satisfaction so these emergency goods can be delivered just in time (Figure 10.8).

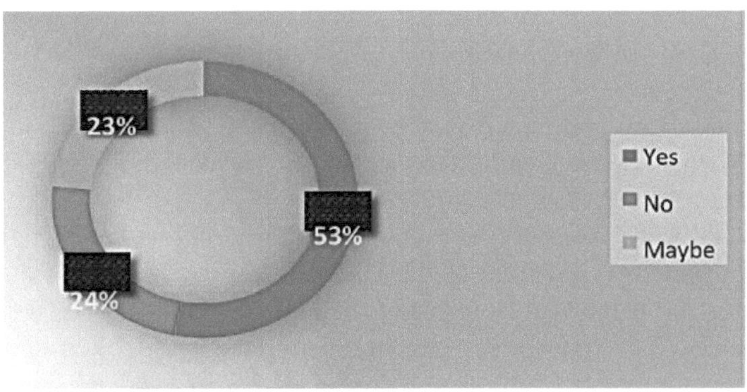

FIGURE 10.8 Preference of online purchase based on quick-need items.

- In order to check the satisfaction level, we asked the subjects if they would prefer buying from platforms that promises the service of Quick Commerce. More the number of candidates agree, higher is the satisfaction level (Figure 10.9).

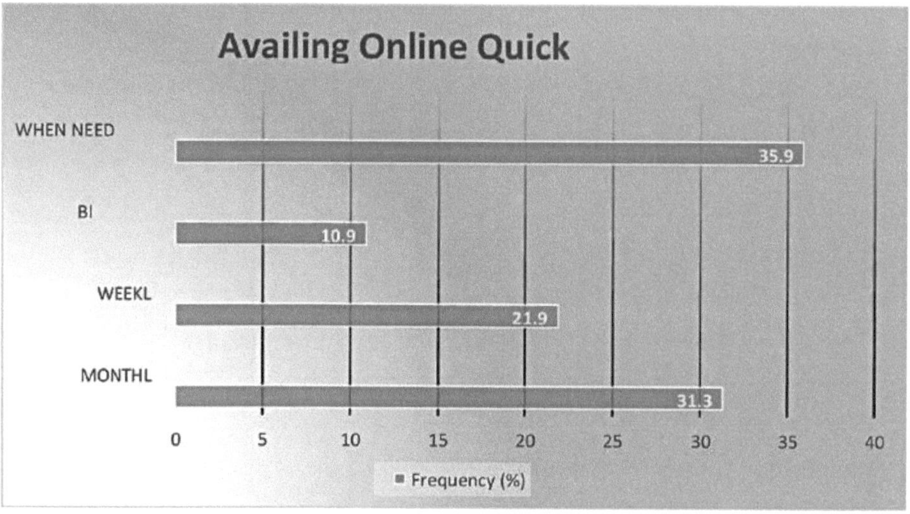

FIGURE 10.9 Frequency of quick-need items purchased by customers: X axis indicates frequency in percentage, Y axis indicates timespan.

Almost 40% of the people are relying on online Quick Services to purchase their products. This clearly explains their trust and dependence on Quick Commerce for those online platforms that already introduced this idea with the customer base.

10.7 RECOMMENDATIONS

Q Commerce is undoubtedly relevant in the 21st century, but there can be challenges in the near future as far as the demand-supply balance is concerned. As per Crunch Base News, people actually prefer "cheaper and faster" delivery. Therefore, if this model can be implemented flawlessly, it can give immense potential to the business houses who will be dealing in eCommerce ("Have You Heard of Q-Commerce? Why And How It's Taking Off," 2022). The main areas that are required to be looked at for proper implementation of Q Commerce are its effectiveness and acceptable service quality. In case faster delivery is promised but is not being fulfilled, then the situation can be cumbersome. Even as per GlobalData, Quick Commerce is something that is forming and developing itself quite rapidly and can act as a favored digital solution, especially whenever there will be a crisis situation (e.g., Lockdown). All that will matter is the customers' experience and whether they are willing to accept this new normal or not ("Q-commerce is rapidly emerging as the next big delivery option, says GlobalData – GlobalData," 2021).

KEYWORDS

- **delivery services**
- **E commerce**
- **online business**
- **Q commerce**
- **sustainability**
- **sustainable marketing**

REFERENCES

Alawadhi, N., (2021). *India Quick Commerce Market to Reach $5 Bln by 2025: RedSeer Report*. Business Standard. Retrieved from: https://www.business-standard.com/article/economy-policy/india-quick-commerce-market-to-reach-5-bln-by-2025-redseer-report-121072600717_1.html (accessed on 6 December 2023).

Bhatia, M., (2022). *How Q-Commerce is Becoming One of the Fastest Growing eCommerce Models?* Indian Retailer. Retrieved from: https://www.indianretailer.com/article/technology/digital-trends/how-q-commerce-is-becoming-one-of-the-fastest-growing-e-commerce-models.a7778 (accessed on 6 December 2023).

Christopher, N., (2022). *Quick Commerce is Reshaping User Behavior in India — and That's a Problem*. South Asia Newsletter. Retrieved from: https://restofworld.org/2022/quick-commerce-is-reshaping-user-behavior-in-india-and-thats-a-problem/ (accessed on 6 December 2023).

E-commerce in India. (2022). Technology, Media & Telecommunications, Deloitte India. Retrieved from https://www2.deloitte.com/in/en/pages/technology-media-and-telecommunications/articles/e-commerce-in-india.html (accessed on 6 December 2023).

Farooq, Q., Fu, P., & Hao, Y., (2019). *A Review of Management and Importance of eCommerce Implementation in Service Delivery of Private Express Enterprises of China*. Sage Open. https://doi.org/https://doi.org/10.1177/2158244018824194.

Global Data Thematic Research, (2022). *Deliveroo and Boots Partnership Sets New Heights for Q-Commerce*. Retail Insight Network. Retrieved from: https://www.retail-insight-network.com/comment/deliveroo-boots-partnership-q-commerce/ (accessed on 6 December 2023).

Gupta, P., Singh, S., Ranjan, R., & Kharayat, G., (2019). *Analysis of Delivery Issues that Customer Face upon eCommerce Shopping* [e-Book]. *International Journal of Management Studies*. Retrieved from: https://www.researchgate.net/publication/333845783_Analysis_of_Delivery_Issues_that_Customer_Face_upon_E-Commerce_Shopping (accessed on 6 December 2023).

Have You Heard Of Q-Commerce? Why and How It's Taking Off. Crunchbase News. (2022). Retrieved from: https://news.crunchbase.com/business/q-commerce-faster-delivery-micro-fulfillment-unicorns/ (accessed on 6 December 2023).

Ma, Q., Chang, C., & Lin, C., (2021). Detecting the crisis of supply chain management on eCommerce for sustainability using q-technique. *Sustainability*. https://doi.org/https://doi.org/10.3390/su13169098.

Morganti, E., Seidel, S., Blanquart, C., Dablanc, L., & Lenz, B., (2014). *The Impact of eCommerce on Final Deliveries: Alternative Parcel Delivery Services in France and Germany* (Vol. 4). Elsevier (Science Direct). https://doi.org/https://doi.org/10.1016/j.trpro.2014.11.014.

Nott, G., (2021). *The 10 Most Powerful People in Rapid Delivery and Q-Commerce*. The Grocer. Retrieved from: https://www.thegrocer.co.uk/technology-and-supply-chain/the-10-most- powerful-people-in-rapid-delivery-and-q-commerce/662483.article (accessed on 6 December 2023).

Propelling India Towards Global Leadership in eCommerce, (2018). NASSCOM. Retrieved from https://nasscom.in/knowledge-center/publications/propelling-india-towards-global-leadership-e-commerce (accessed on 6 December 2023).

Q-commerce is Rapidly Emerging as the Next Big Delivery Option, Says GlobalData – GlobalData. (2021). GlobalData. Retrieved from https://www.globaldata.com/q-commerce-rapidly-emerging-next-big-delivery-option-says-globaldata/ (accessed on 6 December 2023).

Quick Commerce Gets on the Speedy Lane of Growth, (2021). The Hindu. Retrieved from https://www.thehindubusinessline.com/companies/quick- commerce-gets-on-the-speedy-lane-of-growth/article37709939.ece (accessed on 6 December 2023).

Verma, A. M., & Narayanan, C., (2021). *Quick Commerce Gets on the Speedy Lane of Growth*. Thehindubusinessline.com. Retrieved from: https://www.thehindubusinessline.com/companies/quick-commerce-gets- on-the-speedy-lane-of-growth/article37709939.ece (accessed on 6 December 2023).

CHAPTER 11

Key Factors Influencing Customer Preference Regarding Ecotourism from a Data Analytics Perspective

DEBOLINA CHAKRABORTY,[1] AYUSH KUMAR SINGH,[2] KIRTI P. SHARMA,[3] SHIVAM PREM SHARMA,[3] and SWECHA RAM[2]

[1]*Associate Relationship Manager, Standard Chartered Bank, Kolkata, West Bengal, India*

[2]*Future Institute of Technology, Kolkata, West Bengal, India*

[3]*Techno India University, Kolkata, West Bengal, India*

ABSTRACT

With the growth in globalization and technology, individuals now have increased pressure and workload at their workplace. People are putting in hours to triumph in this continuous rat race, which eventually, in most cases, ends up in burnout. Traveling has become essential in this hustle culture. It is the foremost rapidly rising industry, as well as all countries over the globe have an economic benefit attached to it. Collaboration of nations, the influx of foreign currency, and economic boost to tourist zone are some of the advantages of traveling. However, there are several drawbacks too. With more people traveling, the environment gets inversely affected. Therefore, ecotourism is essential now. Adverse effects like harm to the environment, additional pollution, triggering difficulties to the atmosphere, disturbing the ecosystem, and distressing the vegetation, wildlife, and other rare specimens. Travel also surges more exploitation of resources, initiating the scarcity of the same. Overcrowding also intensifies the risks of transmitting dangerous diseases. To diminish the adverse impact of tourism and boost its positive

Sustainability in Marketing Practice: Strategies for Industry 4.0.
Subrata Chattopadhyay, Udit Chawla, (Eds.)

side, the inclusion of proper planning and strict adherence to rules should be followed.

The cities of Kolkata and Howrah have been taken as the data collection point for this study. They are the two most populated and vital cities in the state of West Bengal. They are separated by the river Ganga and connected by two historic bridges and are packed with a number of testaments, monuments, and historical places, making them an attraction for the public eye. Therefore, these cities were the appropriate choice for data collection. This study aims to provide the key parameters that influence an individual's decision and choice regarding ecotourism. Past works by able researchers have been studied and considered to complete this chapter. The four key parameters that are seen reflect customer preference regarding ecotourism. The factors are "subsistence," "traditional heritage," and "conveyance." This chapter intends to analyze the views of various individuals, then further identify the reasons for their estimation, and construct an opinion with the help of SPSS software.

11.1 INTRODUCTION

These days in industrialized nations, specific consideration is given to the expansion of tourism facilities, as it is a vital component of the service industry. Nevertheless, the rapid development of the mass travel industry and ecological inattentiveness have had effects that have directed the demands for the growth of tourism. Ecotourism brings profits to the location, civilization, nation, and economy of desirable zones and resident societies, which offers the finest key to long-standing requirements for protecting and helping the local natural and social diversity. However, a lot of ecotourism growth got flop because of the absence of a good valuation of tourism potential (Lonn et al., 2018).

There is an important method for refining the organizational infrastructure helping ecotourism to grow, the common usage of natural resources, and emerging, refining by controlling outline aimed at ecotourism. Research has consumed attention over defining tourism's influence over the globe, preserving a balance among wildlife, and tourism is emerging as a framework to increase the growth of ecotourism. Simultaneously, a sum of significant chores comprises deciding the effect of environmental tourism on the nationwide economy (Alimov, 2018).

Tourism progress was realized through all individuals involved taking the opportunity to utilize resident natural resources to expand the resident budget, and tourism can become an influential means for the expansion of positivity in the economy on a resident and nationwide measure (Horn & Tahi, 2009; Ecotourism Australia, 2008). Tourism is not only an influential communal and financial power but also a feature of the physical location too. It possesses the power for the advancement of the atmosphere, offers finance for conservation, sanctuary culture, and antiquity set bearable custom bounds and defends natural charms. Ecotourism possibly delivers a sustainable method of progress (Okech, 2009). Ecotourism is a process of natural resource-based travel that is informative, low-impact, non-consumptive, and commonly focused. The resident public should take over the trade and obtain wholesale profits to safeguard sustainable development (D'Angelo et al., 2010). Ecotourism arises from a definitional potential toward encouraging responsible travel to natural areas, to style a positive influence on eco-friendly maintenance, in addition to improving the well-being of residential groups (Angelica et al., 2010; Honey, 2008), and is a tourism type that comprises a topic of preservation of natural parts, teaching, economic improvement, capable tourism, and the contribution of local individuals (Benzer Kilic, 2006).

The significance of understanding the idea of 'ecotourism' - particularly in an emerging country like India - facilitates enhancing neighborhood groups' living and evades distortion of ecotourism purposes. The absence of ecotourism might indicate commercial discrepancy, communal disagreement, and ecological deprivation (Mosammam, Sarrafi, Nia, & Heidari, 2016). Ecotourism provides profits for homegrown people, maintenance support, low-scale expansion, short tourist records, and informative knowledge (Nepal, 2002). Ecotourism has attracted growing devotion in recent years, not only as a substitute for mass tourism but as a source of economic growth and ecological upkeep (Schaller, 2010). Researchers have well-thought-out ecotourism as a key to declining ecological and socio-economic difficulties and as a sustainable expansion instrument in ecologically complex zones. Ecotourism is a vital tool utilized for the input of the protection of the natural scenery and proposes an answer to the deficiency problematic routine in undeveloped areas. Furthermore, it produces a construction efficacy for the economic growth and radical advancement of the resident people, supplying a resource for physical activity of the visitors and the protection (Gregory, 2005; Robert & Santos, 2005; Williams & Ferguson, 2005; Açiksöz et al., 2010).

Due to this possibility, in demand to be effective at and nourish ecotourism happenings which resident public and ecological atmosphere are situated in the center, the primary desire is to identify better standards of individuals and a community environment. Since the regional public is the ones that would disturb and might be affected typically by ecotourism, their contribution and power facility shall be vital. Numerous research shown lately, in what manner the expansions undergone into the tourism takes place apparent through the regional public are studied, and also the behavior of the resident public concerning the outcomes of tourism has been inspected (Kuvan & Akan, 2005; Bertan, 2010; Kiper & Arslan, 2007; Kiper et al., 2009; Mohammadi et al., 2010).

11.2 REVIEW OF LITERATURE

Tourism is a developing enterprise not only in developed countries but also is growing further in primitive countries (Basic et al., 2004). India has various types of natural variations such as mountainous zones, marine beaches, mangrove forests, and many ancient locations that have a massive capacity in the situation of global tourism (Islam, 2006). Ecotourism is measured as a sub-division of ecological tourism and it is meant to have the potential of developing a growing segment of the tourism industry (Lai & Shafer, 2005). Dolnicar & Long (2009) have discussed that ecologically friendly tourists are also willing to search for information about the destination, and whether the destination is ecologically friendly or not. Tourists strive to look for ecological preservation and social protection at the final destination (Buhalis & Law, 2008). Ecological preservation and social protection remain the important pillars of ecotourism (Das & Hussain, 2016); therefore, it specifies the fact that visitors use technology to explore data connected to ecotourism. An inclusive model of ecotourism includes sustainable characteristics which are kind of environmental development, global in scope, and inclusive of both environmental and socio-cultural dimensions (Weaver, 2005).

In the past, main companies have combined ecotourism through cyberspace to distribute details linked to the ecotourism journey end (Das & Hussain, 2016; Lai & Shafer, 2005). So, bases and usages of data have continuously been an arguable matter for the investigators operating in the same research part. Tourists show evidence of a hunting attitude, which allows facility workers to modify their website according to the needs of data pursuers (Zhang et al., 2012). Several groups are designed on the website when different sources of data are positioned for the tourists (Savolainen, 2008). Sites connected to tourism, for example, MakeMyTrip, explore things and items through the internet like Google and also the social networking sites named as Facebook that are used by the customers to gather info (Standing et al., 2014).

Ecotourism has involved growing devotion in fresh years, not individual as an another of mass tourism but also as a process to encourage a nation's financial growth and ecological conservation. It aims toward preserving resources, particularly biological diversity, and uphold maintainable usage of reservations, that can bring ecological involvement to travelers, preserve the eco-friendly environment, and improve (Bansal & Kumaar, 2011; Godretollah et al., 2011; Tewodrose, 2010) Present generation mostly chooses internet for seeking useful data on the problems relating to their everyday natural life goings-on, which is why price efficiency and comfort matter the utmost for them (Buhi et al., 2009). The arrival of technology has transformed the entire procedure of seeking any data counting ecotourism (Standing et al., 2014), and has exclusively permitted travelers. They have also inclined the commercial and profits by the tourism activities. Receiving information to travel is one of the uppermost examined themes on the net, highlighted by Buhalis & Law (2008). Thus, marketers and internet users need to understand that the topic 'ecotourism' has a massive opportunity and ability in today's world (Buhalis & Law, 2008; Buhi et al., 2009).

The method of a permitted agenda for ecotourism is crucial to service native groups and enhance the standard of their existing via employment chances can help in communal participation, social consciousness, organization development, and public consistency. The sustainability of well-planned ecotourism has developed a thoughtful matter in tourism terminus (Surendran & Sekar, 2011). In the definite context, sustainability in tourism turns into an influential reason for marketing schemes and is decisive in an extended bond between providers and clients (Cuculeski et al., 2016). Based on the difference in customer's manners which is more attentive and eager to perform on ecological troubles, life "eco-friendly" develop into a need for businesses and an inescapable price in instruction to continue being competitive in the marketplace (Lewandowska, Witczak, & Kurczewski, 2017). It is rising attention for traveler goods founded on relations by an environment that gives a clear marketplace benefit to ecotourism (Barwicka & Olearnik, 2019). In this view, particular terminuses can take benefit of this propensity, it points to the end that ecotourism should focus on eco-friendly marketing customs in instruction to secure the picture and to stay favored by the visitors.

11.3 OBJECTIVE OF THE STUDY

The objective of the study is to categorize the factors influencing customer preference towards eco-tourism.

11.4 RESEARCH METHODOLOGY

With the use of a structured questionnaire and casual interactions with the consumers, 200 samples were collected from the metro cities of Howrah and Kolkata. The data was collected from people of different age groups and gender. Each parameter studied for the survey was calculated using a five-factor gauge (from 1-strongly disagree, to 5-strongly agree).

11.5 FINDINGS AND ANALYSIS

In Table 11.1, KMO measures of selection capability (which ranges from 0–1), nearer to one are considered well, whereas 0.5 is considered as the minimum necessity. From Table 11.1, the KMO rate is 0.833, which is greater than 0.5 and henceforth, we can continue further the factor analysis. Correspondingly, Bartlett's Test of Sphericity indicates the asset of relationship among those variables. From Table 11.1, we can see that Bartlett's Test of Sphericity is significant, as the significant value less than 0.05 (0.000). Considering the test together, they specify the minimum requirements that must be met before conducting a factor analysis.

TABLE 11.1 KMO and Bartlett's Test

Kaiser Meyer-Olkin Measure of Sampling Adequacy		0.833
Bartlett's Test	Approx. chi-square	2210.053
	df	105
	Sig.	0

Variance explained Table 11.2, every factor expresses a quality score, termed an eigenvalue, under the heading 'Total' of 'Initial Eigenvalues.' Parameters with an eigenvalue of greater than are considered for further study because they only represent true value. It can be seen that Factor 1 (Subsistence) accounts for variance of 7.104 is 47.363% of total variance (Table 11.2), similarly, Factor 2 (Traditional Heritage) accounts for variance of 1.66 is 11.067% of total variance, Factor 3 (Conveyance) accounts for a variance of 1.259 that is 8.396% of total variance, and that's why the original three factor collective account at 66.826.

TABLE 11.2 Total Variance Explained

Component	Initial Eigenvalues			Extraction Sums of Squared Loadings		
	Total	Percentage of Variance (%)	Cumulative	Total	Percentage of Variance (%)	Cumulative
1	7.104	47.363	47.363	7.104	47.363	47.363
2	1.66	11.067	58.43	1.66	11.067	58.43
3	1.259	8.396	66.826	1.259	8.396	66.826
4	0.917	6.111	72.936	–	–	–
5	0.863	5.756	78.692	–	–	–
6	0.65	4.335	83.027	–	–	–
7	0.57	3.798	86.825	–	–	–
8	0.456	3.043	89.868	–	–	–
9	0.415	2.769	92.637	–	–	–
10	0.342	2.283	94.92	–	–	–
11	0.284	1.895	96.815	–	–	–
12	0.225	1.497	98.312	–	–	–
13	0.094	0.628	98.94	–	–	–
14	0.081	0.539	99.479	–	–	–
15	0.078	0.521	100	–	–	–

Note *Extraction method*: Principal component analysis.

Rotated factor matrix (Table 11.3), Factor 1 has a high quantity of variables, the major role played, and benefits provided for the local economy, the development of locals, which in turn promotes the economy. Packing only essentials for a trip, and reuse of items also impact social responsibility. Consequently, this variable is classified "Subsistence." Factor 2 variables indicate high coefficient., travel to places of monumental importance, and cultural heritage is appreciated, and protected to keep it unspoiled. The importance of folk culture is also valued. The understanding of our rich history enhances our quality of living. Consequently, this variable is classified "Traditional Heritage." For the variables, Factor 3 variables indicate high coefficient., traveling using shared vehicles to reduce pollution. Avoiding high fuel emitting options like airplanes, Therefore, this variable may be labeled "conveyance.

$$H0: B1 = 0$$

TABLE 11.3 Rotated Component Matrix[a]

	Component		
	1	2	3
Ecotourism plays an important role in the local economy.	0.914	–	–
Ecotourism policy and planning benefit the local community and environment.	0.906	–	–
Meeting tourists improve the valuable skills of locals.	0.77	–	–
Ecotourism will promote local economic developments.	0.756	–	–
Ecotourism often arises from social responsibility.	0.754	–	–
Packing only what is required is an important step of ecotourism.	0.725	–	–
Reduce–reuse–recycle is essential for ecotourism.	0.673	–	–
Through ecotourism individuals travel to places of monumental importance.	–	0.897	–
Through ecotourism individuals travel to places of cultural heritage.	–	0.804	–
Ecotourism helps protect natural habitats and pristine environments.	–	0.577	–
Ecotourism endorses local folk culture.	–	0.505	–
Ecotourism improves the home-grown quality of life.	–	0.499	–
Ecotourism often involves sharing vehicles with other travelers.	–	–	0.762
Ecotourism involves avoiding travel via airplanes.	–	–	0.641
Ecotourism helps reduce fuel consumption.	–	–	0.522
Extraction method: Principal component analysis.	–	–	–
Rotation method: Varimax with kaiser normalization[a]	–	–	–

[a] *Rotation converged in 6 iterations.*

The null hypothesis implies that there is no line connection between tourist satisfaction and those factors, "subsistence," "traditional heritage," and "conveyance."

$$H1: B1! = 0$$

The alternative hypothesis implies that there is a relationship, positive or negative, between tourist satisfaction and the factors, "subsistence," "traditional heritage," and "conveyance."

Model Summary expresses to what extent a regression model fits the data. In a model immediate R or numerous relationship coefficients measure the value of calculation of the reliant on adjustable. In Table 11.4, R-square rate is 0.499 that indicates 49.9% of total difference in the reliant on variable, complete satisfaction can be described by the independent variables, "subsistence," "traditional heritage," "conveyance."

TABLE 11.4 Model Summary

Model	R	R Square	Adjusted R Square	Std. Error of the Estimate
1	0.706[a]	0.499	0.491	0.507

[a] *Predictors*: (Constant), conveyance, traditional heritage, subsistence.

ANOVA Table 11.5, we can see, the regression model predicts the dependent variable suggestively well and its value is statistically important as the p-value is less than 0.05 (that is 0.000). Thus, H1 is rejected, whereas H0 is accepted.

TABLE 11.5 ANOVA[a]

	Model	Sum of Squares	df	Mean Square	F	Sig.
1	Regression	50.032	3	16.677	65.002	0.000[b]
	Residual	50.288	196	0.257	–	–
	Total	100.32	199			

[a] *Dependent variable*: Tourist satisfaction.

[b] *Predictors*: (Constant), conveyance, traditional heritage, subsistence.

The coefficient Table 11.6 clarifies the complete satisfaction is being derivative from "subsistence," "traditional heritage," and "conveyance," and it can be seen these aspects are statistically significant to form a reversion model or calculation, which is specified below:

$$\text{Customer Satisfaction} = 3.78 + 0.547 * \text{Subsistence} + 0.174 * \text{Traditional Heritage} + 0.411 * \text{Conveyance}$$

TABLE 11.6 Coefficients[a]

Model	Unstandardized Coefficients		Standardized Coefficients	t	Sig.
	B	Std. Error	Beta		
(Constant)	3.78	0.036	–	105.537	0
Subsistence	0.389	0.036	0.547	10.826	0
Traditional heritage	0.124	0.036	0.174	3.44	0.001
Conveyance	0.292	0.036	0.411	8.122	0

[a] *Dependent variable*: Tourist satisfaction.

11.6 CONCLUSION

Tourism is the most essential industry in the service sector of the Indian economy. It is one of the earth's fastest-rising industries and it can take responsibility for accelerating the economic growth of the country and is commonly recognized. It has created several social and profitable benefits, encouraged nationwide incorporation and international awareness, and produced employment opportunities for a large number of individuals and foreign exchange income. Tourism also encourages local handicrafts and cultural performance. Many developing countries, mostly the small countries, are generally reliant upon tourism; this tourism suggests a more dependable source of earnings to them.

This chapter studied the key components responsible for influencing the satisfaction level of the customers regarding ecotourism. It has been found that the customers feel more content with eco-tourism as it has a significant impact on increasing the economy of those locals, helping uplift the local community, and developing their skills. Individuals also wish to travel to places of cultural and monumental importance; it makes them take pride in the country's rich history. Customers are content while traveling using shared vehicles and avoiding polluting ways of travel.

KEYWORDS

- **conservation**
- **consumer preference**
- **ecotourism**
- **principal component analysis**
- **satisfaction**
- **tourism**
- **tourist satisfaction**
- **traditional heritage**

REFERENCES

Açiksöz, S., Görmüş, S., & Karadeniz, N., (2010). Determination of ecotourism potential in national parks: Kure mountains national park, Kastamonu-Bartin, Turkey. *African J. Agric. Res.*, *5*(8), 589–599.

Alimov, A., (2018). *Abstract of the PhD Dissertation on Economic Sciences* (pp. 27–49). Main directions and prospects of development of ecological tourism in the Republic of Karakalpakstan. Samarkand: Zarafshon. Retrieved from http://library.ziyonet.uz/static/lib/readerpdf/web/viewer.html?file=http://library.ziyonet.u z/uplo ads/books/634149/5c0dffe16928e.pdf (accessed on 6 December 2023).

Angelica, M., Zambrano, A., Broadbent, E. N., & Durham, W. H., (2010). Social and environmental effects of ecotourism in the Osa Peninsula of Costa Rica: The Lapa Rios case. *J. Ecotourism, 9*(1), 62–83. Anonymous (2010).

Bansal, P., & Kumar, J., (2011). Ecotourism for community development: A stakeholder's perspective in great Himalayan national park. *International Journal of Social Ecology and Sustainable Development, 2*(2), 31–40.

Barwicka, K., & Olearnik, J., (2019). The behavior of tourism market participants in the context of the development of ecotourism. *Manag. Sci., 24*, 18–25. [CrossRef].

Benzer, K. N., (2006). *The Assessment of the Natural and Cultural Resources of Bolu-Goynuk with the Ecotourism Point of View* (p. 266). Ankara University Graduate School of Natural and Applied Sciences Department of Landscape Architecture, PhD. Thesis.

Bertan, S., (2010). Relationship between tourism support and tourism impacts on socio-cultural in Pamukkale. *Int. J. Econ. Admin. Studies., 2*(4), 83–91.

Buhalis, D., & Law, R., (2008). Progress in information technology and tourism management: 20 years on and 10 years after the internet – the state of e-Tourism research. *Tourism Management, 29*(4), 609–623. https://doi.org/10.1016/j.tourman.2008.01.005.

Buhi, E. R., Daley, E. M., Fuhrmann, H. J., & Smith, S. A., (2009). An observational study of how young people search for online sexual health information. *Journal of American College Health, 58*(2), 101–111. https://doi.org/10.1080/07448480903221236.

Cuculeski, N., Petrovska, I., & Cuculeski, V., (2016). Sustainable marketing and consumers' preferences in tourism. *Eur. J. Tour. Hosp. Recreat., 7*, 84–90. [CrossRef].

D'Angelo, R., Sachasiri, T., Vaccaro, N., Welie, M. V., Vargas, A., & Yongsanguanchai, N., (2010). *Post-Tsunami Ecotourism Development: Solutions for the Laem Khruat Village* (p. 90). Retrieved from: www.wpi.edu/Pubs/E-project/Available (accessed on 6 December 2023).

Das, D., & Hussain, I., (2016). Does ecotourism affect economic welfare? Evidence from Kaziranga National Park, India. *Journal of Ecotourism, 15*(3), 241–260. https://doi.org/10.1080/14724049. 2016.1192180.

Dolnicar, S., & Long, P., (2009). Beyond ecotourism: The environmentally responsible tourist in the general travel experience. *Tourism Analysis, 14*(4), 503–513. https://doi.org/10.3727/108354209X12596287114291.

Ecotourism Australia, (2008). *Climate Action Australia: A Climate Change Response Solution for the Australian Tourism Industry, Sustainable tourism Australia, Australia.*

Gregory, T., (2005). Conflict between global and local land use values in Latvia's Gauja National Park. *Landscape Res., 30*, 413–430.

Honey, M., (2008). *Ecotourism and Sustainable Development: Who Owns Paradise*? Washington, DC: Island Press.

Horn, C., & Tahi, B., (2009). Some cultural and historical factors influencing rural Māori tourism development in New Zealand. *Journal of Rural and Community Development, 4*(1), 84–101. http://report.tuik.gov.tr/reports (accessed on 6 December 2023).

Islam, S. M., (2006). Rural tourism: The right time to contemplate. The daily star (Dhaka), An English daily. *J. Ecol. Natural Environ., 1*(1), 1–6.

Kiper, T., Özdemir, G., & Başaran, B., (2009). Applicability of agricultural tourism and the role of women: A case study of Şarköy Mürefte. *The International Conference of BENA Pollution Management and Environmental Protection* (p. 119). Book of abstract, Tirana, Albinia.

Kuvan, Y., & Akan, P., (2005). Residents' attitudes toward general and forest-related impacts of tourism: The case of Belek, Antalya, *Tourism Management, 26*, 691–706.

Lai, P. H., & Shafer, S., (2005). Marketing ecotourism through the internet: An evaluation of selected ecolodges in Latin America and the Caribbean. *Journal of Ecotourism, 4*(3), 143–160. https://doi. org/10.1080/JET.v4.i3.pg143.

Lewandowska, A., Witczak, J., & Kurczewski, P., (2017). Green marketing today–A mix of trust, consumer participation and life cycle thinking. *Management, 21*, 28–48. [CrossRef].

Lonn, P., Mizoue, N., Ota, T., Kajisa, T., & Yoshida, S. (2018). Evaluating the contribution of community-based ecotourism (CBET) to household income and livelihood changes: A case study of the Chambok CBET program in Cambodia. *Ecological Economics, 151*, 62–69.

Mohammadi, M., Khalifah, Z., & Hosseini, H., (2010). Local people perceptions toward social, economic, and environmental impacts of tourism in Kermanshah (Iran). *Asian Soc. Sci., 6*(11), 1911–2025.

Mosammam, H. M., Sarrafi, M., Nia, J. T., & Heidari, S., (2016). Typology of the ecotourism development approach and an evaluation from the sustainability view: The case of Mazandaran Province, Iran. *Tourism Management Perspectives, 18*, 168–178. doi: 10.1016/j.tmp.2016.03.004.

Nepal, S. K., (2002). Mountain ecotourism and sustainable development, ecology, economics, and ethics. *Mountain Research and Development, 22*, 104–109.

Okech, R. N. (2009). Developing urban ecotourism in Kenyan cities: A sustainable approach. *Journal of Ecology and Natural Environment, 1*(1), 1–6.

Robert, H., & Santos, C., (2005). Tourists and local preferences towards ecotourism development in the Maya Biosphere Reserve, Guatamala. *Environ. Dev. Sustainability, 7*, 303–318.

Savolainen, R., (2008). *Everyday Information Practices: A Social Phenomenological Perspective*. Scarecrow Press.

Schaller, D., (2010). *Indigenous Ecotourism and Sustainable Development: The Case of Río Blanco, Ecuador*. http://www.eduweb.com/schaller/RioBlancoSummary.html (accessed on 6 December 2023).

Standing, C., Taye, J. P. T., & Boyer, M., (2014). The impact of the internet in travel and tourism: A research review 2001–2010. *Journal of Travel & Tourism Marketing, 31*(1), 82–113. https://doi. org/10.1080/10548408.2014.861724.

Surendran, A., & Sekar, C., (2011). A comparative analysis of the socio-economic welfare of dependents of the anamalai tiger reserve (ATR) in India. *Margin: The Journal of Applied Economic Research, 5*(3), 361–379. doi:10.1177/097380101100500304.

Weaver, D. B., (2005). Comprehensive and minimalist dimensions of ecotourism. *Annals of Tourism Research, 32*(2), 439–455.

William, C., & Ferguson, M., (2005). Recovery from crisis: Strategic alternatives for leisure and tourism providers based within a rural economy. *Inter. J. Public Sector Manage., 18*, 350–366.

Zhang, L., Zhu, J., & Liu, Q., (2012). A meta-analysis of mobile commerce adoption and the moderating effect of culture. *Computers in Human Behavior, 28*(5), 1902–1911. https://doi. org/10. 1016/j.chb.2012.05.008.

CHAPTER 12

What Factors Do Zoomers Consider While Buying Private Label Products? A Conceptual Model

DEY SUBHANAN[1] and M. RAJKUMAR[2]

[1]*Research Scholar, Faculty of Management Studies, ICFAI University, Ranchi, Jharkhand, India*

[2]*Associate Professor, Faculty of Management Studies, ICFAI University, Ranchi, Jharkhand, India*

ABSTRACT

The onset of inflation leads to a rise in living costs. The emergence of the private label industry gives customers another option when it comes to shopping. In addition, it gives a sustainable means through which a retailer can survive with their brands over a longer period. Of course, private-label brands offer thicker margins. Because of the rapid progress of PLBs, researchers have looked into the many factors to determine the causes for their success. Research work on private label brands (PLBs) has made tremendous progress in the creation of theory and practical ideas, but the issues of managing private labels at various stages (like low-involvement vis-a-vis high involvement) and demography-wise breakup, especially age-cut, of such individual categories is yet not explored. Although lots have been studied with regards to private-label branding, how is a particular age group like the Gen-Z or Zoomers, behaving about buying private labels, what factors do they consider when choosing a Private Label Brand is something that needs further detailing? Generation-Z or Zoomers are the demographic group that follows the Millennials and comes before Generation Alpha. Mostly, people

Sustainability in Marketing Practice: Strategies for Industry 4.0.
Subrata Chattopadhyay, Udit Chawla, (Eds.)

born between 1995 and 2010s are referred to as Zoomers (Turner, 2015). This generation is the first to have grown up with access to the internet and mobile phones from an early age, so it's no surprise that they're dubbed "Digital Natives," as opposed to the earlier generations who are dubbed as "Digital immigrants." This work aims to close the gap to a great extent by presenting a conceptual model that may be tested later.

12.1 INTRODUCTION

What is it that the future retail marketers need to take care of if they are to sustain in the long run? Which is the Target Group they should run after? What factors should they consider to woo their target audience to them? These are questions that this chapter seeks to answer. Any consumer has a large number of brands to meet their needs, wants, and desires; one of them is the private label brands (PLBs). These are brands that are owned, manufactured, or marketed by the retailers themselves and the distribution of such brands remains exclusively within their retail stores (Valaskova et al., n.d.). PLB, often known as own brands or store brands, are products that a retailer owns, controls, and sells almost exclusively (Baltas & Argouslidis, 2007). Most PLBs have a strong connection to the store, particularly if the company's name is featured on branded items. The assortment of merchandise that the retail chains sell as their brand or private label brand is most often lesser priced than their contemporary manufacturer brands. That is why customers who are more sensitive to price changes or changes in their income are more likely to choose private label products.

PLBs have been successful in the recent past (Redler & Hurth, 2020). Between 2013 and 2018, the dollar volume of PLBs sold by mass retailers increased by 41%, whilst national brands only increased by 7% (Redler & Hurth, 2020). Sarkar et al. (2016) mentioned that every year in the 21st century, the rise of PLBs has exceeded the expansion of manufacturer brands.

The value appraisal as a price-to-value ratio is the most basic element of the private label goods. However, although the price is one factor that a consumer considers before making a purchase decision, the final decision on whether to go in for a private label brand depends on a myriad of factors like product quality, hygiene criteria, and of course the brand name (Valaskova et al., n.d.). India is a young nation, and the majority of the populace is young people. With a median age below China and the other OECD nations, the country has the world's youngest workforce (Jayaprakash et al., 2021). This provides a great opportunity for the marketers to encash on the opportunity, and market their products so that reach across to the target groups like the Millennials and Zoomers (Glynn & Chen, 2009a). Strauss & Howe (1991); Thangavel et al. (2021) suggested the truth that when people grow up in the same period and have comparable socioeconomic experiences, they tend to have similar beliefs, attitudes, and lifestyles. These beliefs shape people's personalities and impact key elements of their lives, such as social interactions and purchasing decisions. However, because the contributing elements and defining events in each geographical region are different, it is hard to generalize the attributes of a generation over the world. Furthermore, the majority of previous studies on generations were conducted in developed nations such as the United States and Europe (Thangavel et al., 2021).

Prior studies on private label products attempted to analyze and identify demographic aspects associated with their consumption. As suggested by Richardson (1997) and Shannon & Mandhachitara (2008), a higher number of members in the family, the more likely they are to buy PLBs. Abramavičienė (2001) reported that higher education was linked to socioeconomic position and quality awareness, both of which appeared to reduce the attraction of private-label products. A systematic literature review (SLR) paper has been done over the past few decades harping on the fact that Private labels are acquiring substantial market dominance over national or manufacturer brands, particularly in the packaged consumer products industry (Ruiz-Real et al., 2017; Riboldazzi et al., 2021). Aside from that, private brands are expanding their market share in developed areas and consolidating their position in emerging markets. Several additional studies have reported inconsistent or contradictory results, such as lesser income is linked to increased price sensitivity, which is linked to a larger interest in private label products (Shannon & Mandhachitara, 2008).

One of the most significant trends in both distribution channel theory and practice is the increase in sales and market share of own brands (Veloutsou et al., 2004), and definitely, one of the most fundamental ways in which retailers have attempted to prosper in recent decades (Bergès-Sennou, 2006). They've turned into a menace to manufacturing brands that has stepped up the process of enhancing research in this particular domain both from the industry and academia point of view (Bao et al., 2011). All the more, rapid changes have been observed in the way the retailers have been operating around the PLBs: retailers are transitioning from being passive channel intermediates to active marketers, introducing innovative store ideas and private brands tailored to certain customer demographics. To compete with the manufacturing brands, a lot of marketing effort is put in, and since the retailers command a "Share-of-shelf," therefore they are in a better position to dictate the eye-level racks and ensure better visibility of products across the retail spaces.

12.1.1 THE SPECIAL CASE OF GEN-Z

It has been seen that Gen-Z and millennials are regularly in touch with social media influencers. Opinion leaders, celebrity endorsers, and social leaders are the three most common categories of social media influencers (Gulamali & Persson, 2017; Hanifawati et al., 2019). According to previous study, these social media influencers have a substantial impact on the views of millennials and Gen-Z, eventually leading to buying behavior (Abreu, 2019) brand switching behavior and change eating behavior (Hanifawati et al., 2019). In the recent past, relevant work in Thailand suggests that it's the influencers that Gen-Z and millennials rely more upon after their friends and family. The most compelling reason to follow influencers was their personality and ability to deliver original, trustworthy, inspirational, and instructional visual material (Hanifawati et al., 2019). They also reported food to be the first and beauty products to be the second preferred category while shopping for PLBs. Work on PLBs conducted in the last five decades has been highlighted by a SLR done by Riboldazzi et al. (2021).

In their chapter, the research gaps had been clearly outlined as – (i) consumer studies with private labels include a comparison of different kinds of merchants, such as by positioning strategy or shop structure; and (ii) value co-creation, Gen-Z customer behavior, social media marketing adoption, are all subjects of private label research. This research chapter seeks to understand what are those antecedents of the final purchase behavior, especially about Gen-Z, a generation whose buying behavior is yet to be ascertained to a large extent. PLBs, being able to generate a larger margin, if marketed well, to the Zoomers will result in greater mileage and sustainability in the long run. Buying behavior of an individual plays a predominant role in consumer behavior in general and among the youth in particular. Buying behavior marketing is the act of building links between market items and certain buying behavior groups. It entails segmenting the market based on purchasing behavior characteristics, presenting the product in a way that appeals to the targeted market's activities, interests, and opinions, and launching targeted promotional efforts.

12.2 REVIEW OF LITERATURE

12.2.1 PRIVATE LABEL BRAND IMAGE

A brand image is defined as the sum of all brand associations in a consumer's memory that contribute to brand perceptions, as cited by Vahie & Paswan (2006). Brand Image is referred to what customers perceive about a brand. It may be described as the customers' view of a certain brand that they have questions in their mind (Bhakuni et al., 2021). It does include the points of similarity and the points of differentiation, committed profits as well as brand personality (Bhakuni et al., 2021). It differs distinctly from Brand Identity in that Brand Identity is what the marketer of a product/service/brand tries to convey to its customers, while the final perception that is imprinted in the brains of the targeted target audience is known as brand image. A great brand image has supported clients to unleash the brand's requirements and helped it to differentiate itself from the competition. Hence, clients must be inclined to go for the said brand (Dam & Dam, 2021). These types of companies indeed enjoy a more favorable position as compared to their competitors, have a sustainable competitive advantage, and increase their market share as well as performance (Dam & Dam, 2021). Store Image, following similar lines, is a concept that has been of fundamental importance both to academic researchers and industry practitioners (Kumar et al., 2014; Mostafa & Elseidi, 2018).

The studies conducted by Malaysia et al. (2018) have been able to substantiate one particular thing – a store image is a collection of traits (both real and imagined) that customers associate with a certain retail establishment. In line with the aforementioned, Beneke et al. (2012) defined store image as a collection of commercial brand associations that belong to the outlet in the consumer's mind as a result, the aforementioned brand image associations are multidimensional, consisting of the emotive dimension, or sentiments toward the brand, as well as the perceived quality component (Vahie & Paswan, 2006). Store Image may also be put as "how the store is defined in the shopper's mind partly by its functional qualities and partly by an aura of psychological attribute" (Retailing: Critical Concepts, n.d.). As has been explained by Kumar & Kaushal (2017), during the study of brand loyalty in the Indian terrain, single-brand loyalty is no more the order of the day, and only dual or multi-brand loyalty has replaced the earlier concept. Adi et al. (2020) pointed out that there is an interplay between the reputation of a brand and the functional and affective image of the brand image. A positive image of a brand goes on to enhance loyalty toward the brand.

12.2.2 IN-STORE EXTRINSIC CUES

One of the major theories that are referred to by researchers is the Cue Utilization Theory. This theory establishes a logical framework for evaluating customer perceptions on the quality of the store brand (Dick et al., 1995). Mostly, before making a purchasing choice, buyers are uninformed of the true quality of rival brands' products. Consumers are more inclined to use basic heuristics, or cues, to judge product quality in such instances, according to study. The bulk of these signals are product market aspects (Dick & Richardson, 1997). Extrinsic signals are those that are more noticeable than intrinsic clues. Since consumers may not have access to the product's intrinsic cues or have insufficient information about the items, consumers are more likely to rely on extrinsic criteria when evaluating product quality (Brouthers & Xu, 2002). Customers seldom have the time or desire to examine items extensively before making a purchase in many circumstances, especially for some categories in concern, such as low-involvement FMCG products. It is then that from their memory, a set of cues in assessing the product quality arise, and they arise by the predictive or confidence (CONF) values (Richardson et al., 1994). Predictive value refers to the degree to which customers identify a specific indication with product quality. This is analogous to a cue's diagnostic capacity, which relates to how reliable a cue is in distinguishing between different categorizations. As a result, the more diagnostic the signal, the more likely it is to be employed in the evaluation of product quality (Dick & Richardson, 1997). The CONF value of a cue refers to the degree to which consumers are confident in their ability to utilize and assess that cue appropriately (Dick et al., 1996).

In the framework given by Richardson et al. (1994), It is advocated that merchants use extrinsic high scope cues to boost consumer perceptions of PLBs' quality. Extrinsic cues appear to be more easily identified, integrated, and understood, and hence are more likely to have a higher CONF value (Richardson et al., 1994). Similarly, signals with a broad reach are regarded as more credible and have a higher predictive value. As a result, broad usage of extrinsic high scope value will result in high predictive value and high CONF value, significantly increasing store brand quality judgments (Dick et al., 1995). There could be certain contributors to these extrinsic cues, like Packaging, Store-based promotions, Home Delivery.

12.2.2.1 PACKAGING

Extrinsic cues include packaging. According to certain scientific findings from a virtual reality simulation, package images appear to attract a shopper's attention to a specific brand. However, as per the study conducted by Underwood et al. (2001) suggests that this effect is contingent, as this is prevalent only for the low-familiarity brands in the categories that are offering a high level of experiential benefits. Subsequently, therefore packaging needs to be improved to enhance the overall likeability of the brand (Underwood et al., 2001). There is evidence that brand names and packaging have a significant impact on quality perception. When the simplicity of evaluation is minimal, brand reputation can be used as a common surrogate indication of product quality or as an effective risk management approach (Oosthuizen, 2015).

12.2.2.2 STORE-BASED PROMOTIONS

Big Bazaar, Pantaloon Retail's hypermarket, released a morning cereal line under its own label, Tasty Treat, in 2013. Soups, carbonated drinks, noodles, pickles, nankeens, ketchup, chips, jams, bread, Khari, vermicelli, papads, and fruit beverages were already available under the Big Bazaar name. It was then trying to ass breakfast cereals to the range (Iyer, 2015). They tried out a lot of in-store promotions, especially, a free sampling of the corn flakes in the category area, which saw a temporary rise in business numbers, especially on the Wednesday Bazaars when the in-store promotions were on in full swing. The conclusion is that merchants may utilize free samples to improve consumers' perceptions of product quality by allowing them to assess things based on intrinsic rather than external cues. Retailers may also employ a variety of extrinsic signals in visual merchandising to create the correct esthetic impression surrounding their PLBs (Maharani et al., 2020; Mishra et al., 2020).

12.2.2.3 HOME DELIVERY

Customers have diverse needs today, and of prime importance is Home Delivery to segment customer experience and ease. Therefore, retailers could assure free home delivery once they are buying private label products.

12.2.3 PERSONALITY FACTORS

The psychological systems, as well as an individual's unique pattern of thinking, emotion, and conduct is referred to as Personality traits or factors

(Funder, 1997; Herstein et al., 2012). Therefore, they are a major factor in explaining human tendencies behind a particular action, all the more in the case of Private label brand purchases. While personality should be an indicator to predict job performance, marriage stability, and even mortality (Herstein et al., 2012), There is evidence of the importance of personality in psychological research, but there have been few studies of its impact on consumer behavior, particularly in the case of PLBs among different age groups.

Although they are classified as cultural characteristics, they may also be viewed as individual personality traits whose diversity can and should be researched across cultures (Beneke et al., 2012). In the grocery category, personality traits such as risk aversion, value consciousness, deal proneness, brand loyalty, price-quality judgments and responsiveness to promoted discount products, coupon usage, were investigated (Mishra et al., 2020). Value consciousness and Deal proneness determine the private-label purchase.

Some of the underlying pointers are:

1. **Individualism:** It is founded on the idea that people are self-contained and independent of one another (Herstein et al., 2012). Also, as per Herstein et al. (2012), Individualism must be seen as a combination of a focus on rights above obligations, care for oneself and one's immediate family or dependents, emphasis on personal autonomy and self-fulfillment, and basing one's identity on one's achievements (Herstein et al., 2012). Hitherto, however, the maiden study to research comparatively between the individualistic and collective personas and the brand choice was conducted (Herstein et al., 2012).

2. **Materialism:** It is defined as an enduring belief in the value of acquiring and possessing things, with three major components: "acquisition centrality," the role that said acquisition plays in bringing happiness, and the role of possessions in defining success. According to cross-cultural research, this is a significantly more common behavior in the Western world, and evidence reveals that customers in less economically developed countries wish to copy their counterparts in more sophisticated economies with whom they engage.

3. **Need for Cognition:** A person with a higher need for cognition is more likely to put out substantial cognitive effort and elaborate on

the information provided, whereas someone with a lesser need for cognition is more likely to avoid thinking (Herstein et al., 2012) Because they are innately interested in analyzing, digesting discrete pieces of information, and like thinking about product-related information, persons with a strong demand for cognition are prone to assessing items based on the strength and relevance of product attribute arguments. Those who have less needs and are more driven by the final result rather than the procedures will be less motivated to investigate the data in front of them. They are, on the other hand, more likely to make decisions based on criteria such as country of origin. Despite substantial research in the field of psychology on the relevance of cognition, there are no similar contributions in the field of consumer behavior.

12.3 RESEARCH PROPOSITION AND THEORETICAL MODEL

12.3.1 PRIVATE LABEL BRAND IMAGE AND PERCEIVED QUALITY OF PRIVATE LABEL BRAND

Consumers pick their brand or products based on the expected brand image; hence consumer perceptions of brand image have a favorable impact on purchase intent (Cheng et al., 2007). Apart from that, brands continue to play an essential role in buying decisions, as a positive brand image will reflect on product quality, resulting in an increase in purchase intent (Ramadhan & Muthohar, 2019). Aside, a positive brand image will lessen consumer perceptions of purchasing risk (Cheng et al., 2007). Consumers try and evaluate the brand image of the Private Label in terms of several aspects like the merchandise quality, the kind of advertisements both ATL and BTL, positive word of mouth that spreads amongst the people they mingle with, and often Co-Creation, which is very much intrinsic to Gen-Z (Abramavičienė, 2001). As cited by Ramadhan & Muthohar (2019) retail stores do compete with the manufacturer brands and try to provide better quality merchandise as compared to similar counterparts of manufacturer brands at a much cheaper rate. However, because the private brand price is cheaper, it has a simpler packaging and usually is lesser known since it is not much advertised. As a result, the consumer believes that private brand items are riskier due to the low quality of those products (Ramadhan & Muthohar, 2019). Ailawadi (2021) also thinks that PLBs are often thought to be of poorer quality than manufacturer brands since private label products do not adhere to international packaging standards. The Store Image and the Private Label Brand

Image are commonly confused., As a result, the stronger the store's image is, the more favorable brand association the private labels it house have.

According to current research, store image is defined as a collection of extrinsic features such as store name, goods quality, competent salespeople and store (Mostafa & Elseidi, 2018). On the other hand, perceived quality is a consumer's evaluation of a product's overall excellence or superiority based on its inherent and external attributes (Beneke et al., 2012). The probable effect going forward depends on the situational and personal factors (Beneke et al., 2012). As a result, the perceptual qualification is included. As a brand's perceived quality rises or falls, so does its perceived prestige, as cited by Beneke & Carter (2015a). The higher intensity of distribution gives the manufacturer brands an advantage with regards to the extrinsic cues over the PLBs, in the sense, that its distribution is restricted only to the stores. Consumers who are brand loyal are more likely to purchase the specific manufacturer brand in discussion merely because of the emotional tie and happiness they had during past purchases (Glynn & Chen, 2009a). Dick et al. (1995) stated that consumers consider store names as one of the primary indicators of PLB quality.

- **Proposition 1:** Private label brand image influences perceived quality of the private label product.

12.3.2 IN-STORE EXTRINSIC CUES AND PERCEIVED QUALITY OF PRIVATE LABEL BRAND

Private label brands (PLBs) are the intellectual property that is either developed or marketed or both by the retail groups. The fact that the retailer's extrinsic cues are likely to reflect significantly on their PLBs, with some consumers even considering these brands as extensions of the retail brand itself, is worth noting (Collins-Dodd & Zaichkowsky, 1999). Vahie & Paswan (2006) advocated that there is a significant correlation between the store's image, customer knowledge, and the PLB relationship. Consumers make many selections on-the-spot, hence research has previously emphasized the relevance of within-the-store (Mishra et al., 2020). In-store marketing is an essential marketing technique for Private Label businesses, according to Beneke et al. (2012). According to earlier study referenced by, in-store marketing, such as shelf space management, notably at gondola ends and category gondolas, displays, posters, and in-store communication, has a positive influence on brand attention and sales (Mishra et al., 2020). Directional signage, point-of-purchase signage, and images are all useful techniques for raising PLB knowledge and attitude. Retailers are interested in promoting in-store brands not only because of the higher profits, but also because it reduces dependency on national brand manufacturers, improves in-store image, and increases consumer loyalty (Mishra et al., 2020). Despite this, retailers face a significant problem in persuading consumers to switch from national brands to private label goods. PLBs were positioned as an economical alternative to major brands in the early years of the private label boom. The situation is changing, however, as customers are increasingly demonstrating an interest in PLBs (Mishra et al., 2020). Retailers, unlike NBs, are unable to spend considerable resources to mass media advertising of their PLBs. In such a setting, in-store marketing campaigns can play a critical role in fostering consumer acceptance of PLBs (Mishra et al., 2020). Packaging, store-based advertising, and home delivery are examples of extrinsic signals that have been explored in previous sections.

- **Proposition 2:** Extrinsic cues, influences the perceived value of the private label brand.

12.3.3 PERSONALITY AND PERCEIVED QUALITY OF PRIVATE LABEL BRAND

Personality is most often ascribed to traits like self-confidence, dominance, etc. It's a unique psychological element that may tend to have a relatively consistent and lasting response to an individual's environment. Consumers choose brands that have personalities that are similar to their own. As a result, a brand personality is made up of the distinctive human traits that may be attributed to a single brand. The self-congruence hypothesis of consumer behavior demonstrates that there is a link between personality and product choice (Erdoğmuş & Büdeyri-Turan, 2012). Personality does have a significant impact on brand selection and decision-making. Consumers prefer, become loyal to, and advocate companies and goods that reflect their current or ideal self-concepts (Erdoğmuş & Büdeyri-Turan, 2012). Because the ready-to-wear market has higher levels of conspicuousness, symbolism, and product engagement, the relationship is anticipated to be publicly communicated, meaning that personality provides way to convince a consumer to choose a certain brand or not (Mishra et al., 2020). Consumers who exude higher levels of individualism, materialism, and a low need for cognition were the ones who later placed great importance on brand image dimensions like package design, country of origin, and reputation of the national label brand, which are all extrinsic, as described in the earlier section on personality. Materialistic consumers like to brag about their better lifestyles to their peers and to themselves. Tifferet & Herstein (2010) stated that Consumers with a lesser cognitive need view the brand to be of higher quality. Customers who rely on such indications are more likely to select well-known brands than those who do not (Herstein et al., 2012).

- **Proposition 3:** Personality factors will be positively correlated with the perceived quality of the PLB.

12.3.4 EFFECT ON PERCEIVED VALUE FOR MONEY

Positive perception for the PLBs is created by the Perceived value that the consumer has for the same (Gangwani et al., 2020a). The consequence or effect of a brand's purchase and consuming experiences is what is referred to as perceived value (Gangwani et al., 2020a). The term "perceived value for money" refers to how customers feel about a product's pricing (Gangwani et al., 2020b). The perceived worth of a brand, in relation to its price, comprises a perception of brand quality, but not in its whole (Dick et al., 1996). Perceived value is described as "that which is regarded to be fair, proper, or merited in proportion to the perceived cost of the product, while taking into account relevant competitor alternatives," according to equity theory terminology (Gangwani et al., 2020b). In the setting of a retail store, perceived value is proven to be one of the antecedents of purchase intentions. The most important factor of purchase intention is, without a doubt, the consumer's perception of value (Gangwani et al., 2020a).

Private label buyers are price-conscious, but they still want good value for their money. Consumers benefit from own-label products since they are more affordable and of higher quality. As has been noticed, higher-quality brands instill CONF in customers (Mostafa & Elseidi, 2018). Apart from that, Richardson et al. (1994) claim that customers' perceptions of Private label brand quality affect its purchase volume and market share. But Mostafa & Elseidi (2018) revealed through their study that the buying of private label products by quality-conscious Thai customers showed a negligible correlation. In consumer behavior, the link between customer perception and attitude has long been recognized. Consumer attitudes and perceived quality have a significant positive association (Richardson et al., 1994).

- **Proposition 4:** Perceived quality is positively correlated to perceived value for money.

12.3.5 PRIVATE LABEL ATTITUDE

Attitude is emphasized in several consumer behavior models, and it serves as the foundation for effective segmentation, market forecasting, and marketing program development (Binninger, 2008). Abramavičienė (2001) defines attitude as a positive or negative propensity toward the qualities of a certain thing, which influences the customers' tendency to behave predictably toward that object.

12.3.6 PERCEIVED RISK IN BUYING PRIVATE LABEL BRANDS (PLBS)

"Perceived Risk" refers to a consumer's skepticism about the outcomes of their purchasing decisions. The various types of perceived risks are as under:

1. **Financial Risk:** The chance of losing money as a result of a poor buying decision or by purchasing an unknown or unfamiliar brand. It would also involve the danger that the product's quality isn't up to par with the amount paid, or that it's overpriced and could have been bought for less elsewhere.
2. **Social Risk:** A potential loss of image or status as a result of purchasing a specific brand or product.
3. **Physical Risk:** The customer is concerned that purchasing a product or brand may result in a decline in health; in other words, health may be harmed as a result of purchasing a certain brand.
4. **Functional (Performance Risk):** The fear that the outcome of a goods purchase will not satisfy customer expectations. It is the concern that the consumer's purpose for purchasing a product will not be met by the product or brand in issue.
5. **Psychological Risk:** The consumer may be disappointed after choosing a bad product or service pick (Arslan et al., 2013).

Financial risk would be included in studies on purchase risk at the product category level, social risk, and financial risk, as ascertained by Horvat & Ozretić Došen (2013).

- **Proposition 5:** Lower perceived risk will lead to a better attitude towards the purchase of private label brands.

12.3.7 PERCEIVED RELATIVE PRICE OF PRIVATE LABEL BRANDS (PLBS)

Price-conscious customers are less likely to be brand loyal, according to previous studies, and instead demonstrate strong variety-seeking behavior (Glynn & Chen, 2009a). Those that buy PLBs on a regular basis have a similar pattern of activity. Consumers who do not have a desire for a consistent brand are those who fall into this category. When the price difference between the national and private brands narrows, they may simply move to the manufacturer brands. The likelihood of obtaining a PLB is inversely related to brand loyalty (Glynn & Chen, 2009a). As cited by Mishra et al. (2020), The term “price consciousness” refers to “buyer’s unwillingness to pay a higher price for a product.” The cheaper pricing of PLBs compared to national brands are the reason why they are purchased (Glynn & Chen, 2009b) Furthermore, retailers are encouraged to design and offer superior in-store placement, pricing, and displays to entice customers to choose PLBs over NBs. PLBs are preferred by consumers who are prepared to pay modest rates and have a good attitude about them (Ailawadi, 2021). According to Menon (2018), PC has a direct beneficial relationship with the buying intention of fashion companies in the fashion business. Research conducted earlier in a similar domain, like the one done by Munusamy and Wong, as stated by Cela et al. (2016), suggested that there is considerable positive association between perceived price and consumers’ purchase intentions towards PLBs. Overall, the literature indicates that PLB purchase increases when there is a higher perceived relative price of the PLB.

- **Proposition 6:** Higher the perceived relative price of the private label brand, the better would be the attitude towards buying the same.

12.3.8 PERCEIVED VALUE FOR MONEY AND ITS EFFECT ON THE ATTITUDE OF BUYING PRIVATE LABEL BRANDS (PLBS)

Perceived value infers to the mental evaluation in the minds of customers relating to specific goods or services (Yang & Peterson, 2004; Beneke & Carter, 2015b). Individuals do not buy products or services based only on the functional benefit but on the value, they derive (Beneke & Carter, 2015b). Value-attitude and behavior (VAB theory) proposed by Homer & Kahle (1988) theorized the flow from value to the attitude of the consumers

(Ghali-Zinoubi, 2021). In case the perceived value for money is higher, then it has a positive influence on buying of private-label brands.

- **Proposition 7:** Higher the perceived value for money, the greater will be the positive attitude toward private label brands.

12.3.9 EFFECT OF ATTITUDE ON PURCHASE INTENTION

Purchase Intention is defined as the likelihood that customers will intend or be willing to purchase a specific product, brand, or service in the future, and it is the last stage before engaging in real purchasing action (Kakkos et al., 2015). Purchase intention measures examine and quantify future purchasing behavior. They're cheap, simple to understand, and interpret, hence they're commonly employed in sales forecasting. These statistics are commonly used to forecast demand for new items, impacting production schedule regulation, advertising, distribution, and price policy on an operational level (Kakkos et al., 2015). To figure out what are the drivers of intentions while the purchase of PLBs, one needs to take into account firstly that buying decisions can be influenced by various needs including physiological and/or socio-psychological (e.g., prestige, recognition, comfort) to then, focus on factors likely to form customer attitudes towards PL branded products.

- **Proposition 8:** Greater the attitude, the better would be the intention towards buying private label brands.

12.4 THE CONCEPTUAL MODEL

12.4.1 THE THEORY OF REASONED ACTION AND PLANNED BEHAVIOR

The Theory of Reasoned Action and Planned Behavior serves as the foundation for the conceptual model. In 1980, the Theory of Reasoned Action was developed to forecast an individual's intention to participate in a certain behavior at a specific time and location. The idea attempted to identify all actions that humans may exert self-control over. Behavioral intent is a key component of the model, and it is impacted by one's attitude about the likelihood that the activity will produce the desired results, as well as one's subjective assessment of the risks and advantages of that outcome.

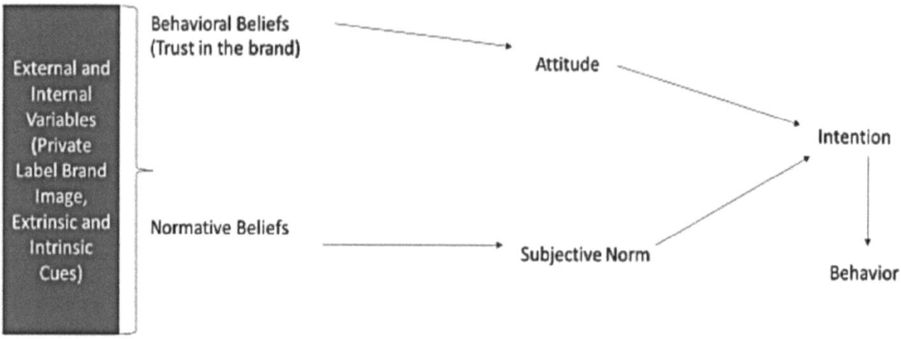

FIGURE 12.1

Source: Authors

1. **Attitudes:** These are described as a person's positive or negative assessment of a certain activity of interest. It requires thinking about the consequences of executing the activity.
2. **Behavioral Intention:** The motivating elements that impact a certain behavior, with the greater the desire to conduct the activity, the more likely it will be accomplished.
3. **Subjective Norms:** This is the notion that the majority of people accept or disapprove of a particular conduct. It has to do with a person's ideas about whether or not peers and important individuals in his or her life think he or she should partake in the conduct.
4. **Social Norms:** A group of people's or a wider cultural context's conventional regulations of behavior. In a group of individuals, social standards are regarded normative, or standard.
5. **Perceived Power:** It refers to the perceived presence of factors that may facilitate or impede the performance of a behavior. Perceived power contributes to a person's perceived behavioral control over each of those factors.
6. **Perceived Behavioral Control:** This relates to an individual's perception of how easy or difficult it is to do the desired action. Perceived behavioral control fluctuates according to events and acts, resulting in a person's sense of behavioral control changing depending on the scenario. The theory of planned behavior (TPB)

was born out of the move from the theory of reasoned action to the TPB (The Theory of Planned Behavior, n.d.) (Skewes & Gonzalez, 2013).

12.4.2 THE CUE UTILIZATION THEORY

The cue utilization theory is based on the fact the presence of the cue raises the marginal utility one derives from consumption. Whenever any prospective consumer makes a selection of any particular product, they essentially go in for the selection of a particular cue in question, which could be "The Perceived Value or the Convenience value" or "The Intrinsic and extrinsic cue"; to make a comprehension of the same; their beliefs eventually have a direct impact on the way they comprehend the values and the decision made thereof is based on the combined effect of the two. Coming to the choice of private-label brands, cue utilization theory is made used to derive customer perceptions about store brand quality (Dick & Richardson, 1997). Most academics utilize this theory to explain how consumers perceive quality for various cues. According to this idea, each given product is made up of a variety of cues that customers use as quality indicators (Dick & Richardson, 1997). It's also been noticed in several cases, the shoppers themselves do not know of product form or product category competitors while they are making their decisions; and it is here, that they majorly depend on simple heuristics (referred to as "cues") to assess the product quality. There is also something called "Impulse products" and herein, prospective customers neither have the time nor the bait to compare products; thus, all that they have is simple heuristics to assess product quality (Srinivasan & Till, 2002).

Cues can have a couple of dimensions. Cues can be characterized as either extrinsic or intrinsic to the products at first. Intrinsic cues are embedded in the product and, if altered, will result in changes to the product's makeup, such as components or ingredients. Extrinsic cues are product features that are not part of the physical product and may be modified without changing the product's composition, such as price and brand name (Gamliel & Herstein, 2007). The feature that distinguishes extrinsic from intrinsic is the fact that while the former is high on the CONF value, the latter is high on the predictive value. Second, several studies have divided the cues into two categories: high scope and low scope cues.

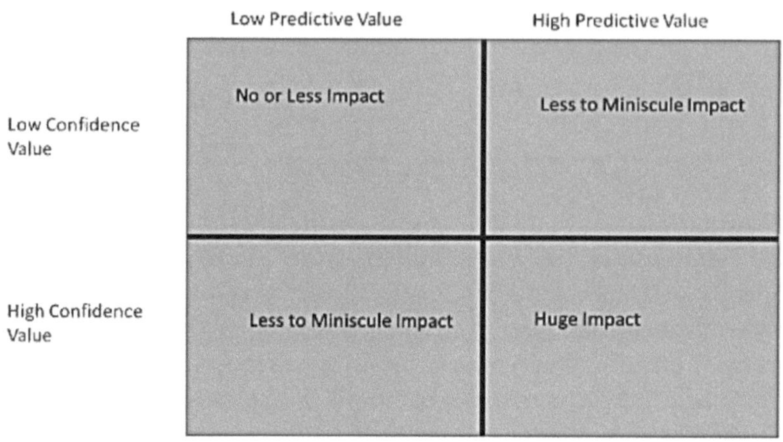

FIGURE 12.2

Source: Authors

Thus, now in the first place, the cue utilization theory can be taken as a theoretical model on which further research can be done.

12.4.3 YOUNG AND RUBICAM'S BRAND ASSET VALUATOR MODEL

As per Young & Rubicam (2000), examining the interactions between people and brands is akin to studying a brand. The four pillars that make up a brand are differentiation, relevance, esteem, and knowledge.

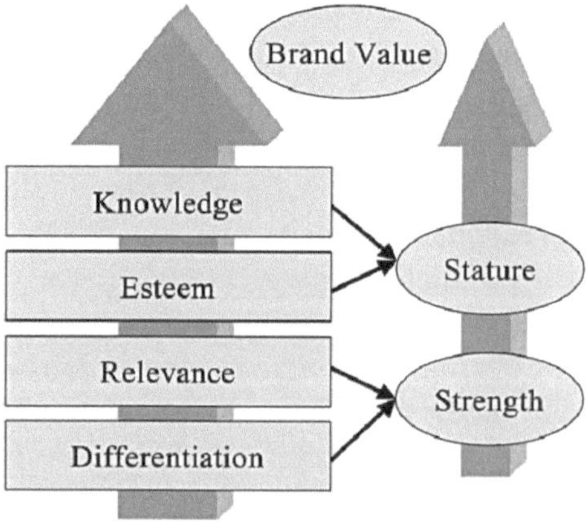

FIGURE 12.3

Source: Reprinted with permission from de Mortanges & Van Riel (2003).

Differentiation is innate in brands, and it is one of the most important factors in their success. Consumers will be more interested in learning more about a brand if it is differentiated. Even during the mature period of the product life cycle, brands will struggle incessantly to preserve their distinction, despite the fact that differentiators are swiftly exploited by competitors. To symbolize its relevance and guarantee that the brand is sustaining and reaching the market, the level of distinctiveness should be very distinctive. Because of the high level of brand rivalry, developed-country brands should be more distinct than those in developing and less developed nations. In order to achieve brand margin in the market, the brand's relevance will become the driving force. Consumers' willingness and devotion to a specific brand is measured by brand esteem. Brand esteem normally develops once distinctiveness and market relevance have remained stable. Certain brands, on the other hand, may rely purely on their reputation if they have been on the market for a long period. Brand esteem, on the other hand, is the sum of quality and popularity based on experience of the consumers. They will take time to learn about a brand, and it is generally the end consequence of a company's marketing efforts. Young & Rubicam (2000) established the Brand Asset Valuator (BAV) model, which consists of two key components that make up a brand: (i) brand strength, which causes a brand to lead in the market; and (ii) brand stature, which causes a brand to lag in the market (Hassan & Rahman, 2012).

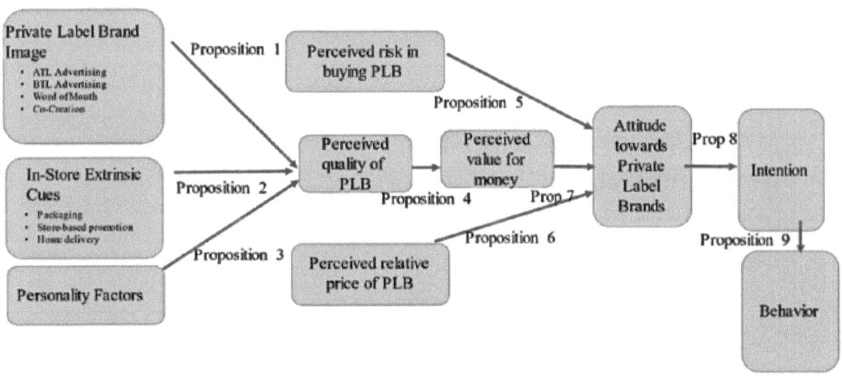

FIGURE 12.4 Conceptual Model as developed by authors

12.5 IMPLICATIONS AND CONCLUSION

Private label brands (PLBs) are taking all the efforts to be on a level with the growing market share that the PLBs are capturing. Various factors interplay while the consumer is trying to purchase the PLBs. This chapter seeks to understand those major factors that drive consumers, especially, the Gen-Z consumers who are going to form a huge gamut of the population trying to barge into buying PLBs. This study will add to the body of knowledge on customer behavior by emphasizing the relevance of Gen-Z and Private Label branding. To summarize, the Zoomers or Gen-Z have a bigger effect on individual and family purchasing behavior due to their purchasing behavior, purchasing power, awareness, and interest in items. Since Gen-Z accounts for a significant share of the Indian market, every marketer must understand the mindset of this generation in order to differentiate themselves from the competitors in the future. Changing buying behavior makes it all the more important for marketers to craft their communication strategies and mediums of communication accordingly to suit the needs of Gen-Z. Such a customer-oriented marketing approach would woo the Zoomers and help in sustenance in the long run.

KEYWORDS

- **category characteristics**
- **conceptual model**
- **consumer characteristics**
- **emerging market**
- **generation Z**
- **in-store extrinsic cues**
- **personality factors**
- **private label brands**
- **Zoomers**

REFERENCES

Abramavičienė, R., (2001). *Consumer and Image Factors Affecting Private Label Purchase of Major Lithuanian Retailers: Maxima." Iki" and" Rimi.* Jin and Suh.

Adi, A., Merima, C., & Emir, A., (2020). The role of brand image in consumer-brand relationships: Similarities and differences between national and private label brands. *Management & Marketing. Challenges for the Knowledge Society, 15*(1), 1–16.

Ailawadi, K. L., (2021). Commentary: Omnichannel from a manufacturer's perspective. In: *Journal of Marketing* (Vol. 85, No. 1, pp. 121–125). SAGE Publications Ltd.

Baltas, G., & Argouslidis, P. C., (2007). Consumer characteristics and demand for store brands. *International Journal of Retail & Distribution Management, 35*(5), 328–341.

Bao, Y., (2004). Association for consumer research discerning store brand users from value consciousness consumers: The role of prestige sensitivity and need for cognition discerning store brand users from value consciousness consumers: The role of prestige sensitivity and need for cognition. *Advances in Consumer Research, 31.*

Bao, Y., Bao, Y., & Sheng, S., (2011). Motivating purchase of private brands: Effects of store image, product signatureness, and quality variation. *Journal of Business Research, 64*(2), 220–226.

Be Beneke, J., & Zimmerman, N. (2014). Beyond private label panache: the effect of store image and perceived price on brand prestige. *Journal of Consumer Marketing, 31*(4).

Beneke, J., & Carter, S., (2015). The development of a consumer value proposition of private label brands and the application thereof in a South African retail context. *Journal of Retailing and Consumer Services, 25*, 22–35.

Beneke, J., Greene, A., Lok, I., & Mallett, K., (2012). The influence of perceived risk on purchase intent – the case of premium grocery private label brands in South Africa. *Journal of Product & Brand Management, 21*(1), 4–14.

Bergès-Sennou, F., (2006). Store loyalty, bargaining power and the private label production issue. *European Review of Agricultural Economics, 33*(3), 315–335.

Bhakuni, P., Rajput, S., Sharma, B. K., & Bhakar, S. S., (2021). Relationship between brand image and store image as drivers of repurchase intention in apparel stores. *Gurukul Business Review, 17*(1).

Binninger, A. S., (2008). Exploring the relationships between retail brands and consumer store loyalty. *International Journal of Retail & Distribution Management, 36*(2), 94–110.

Brouthers, L. E., & Xu, K., (2002). Product stereotypes, strategy, and performance satisfaction: The case of Chinese exporters. *Journal of International Business Studies, 33*(4), 657–677.

Cela, S., & Cazacu, S. (2016). The attitudes and purchase intentions towards private label products, in the context of economic crisis: a study of the Salonian consumers. *Ecoforum Journal, 5*(3).

Collins-Dodd, C., & Zaichkowsky, J. L., (1999). National brand responses to brand imitation: Retailers versus other manufacturers. *Journal of Product & Brand Management, 8*(2), 96–105.

Dam, S. M., & Dam, T. C., (2021). Relationships between service quality, brand image, customer satisfaction, and customer loyalty. *The Journal of Asian Finance, Economics and Business, 8*(3), 585–593.

De Mortanges, C. P., & Van, R. A. C. R., (2003). Brand equity and shareholder value. *European Management Journal, 21*(4), 521–527.

Dick, A., & Richardson, P., (1997). How consumers evaluate. *Pricing Strategy & Practice, 5*(1), 18–24.

Dick, A., Jain, A., & Richardson, P., (1995). Correlates private brand proneness. *Journal of Product & Brand Management, 4*(4), 15–22.

Dick, A., Jain, A., & Richardson, P., (1996). *Dick 1996, 5*(2), 19–28.

Erdoğmuş, I., & Büdeyri-Turan, I., (2012). The role of personality congruence, perceived quality, and prestige on ready-to-wear brand loyalty. *Journal of Fashion Marketing and Management, 16*(4), 399–417.

Evaluating Website Quality: Applying Cue Utilization Theory to WebQual | Semantic Scholar. (n.d.). Retrieved from: https://www.semanticscholar.org/paper/Evaluating-Website-Quality%3A-Applying-Cue-Theory-to-Longstreet/14066a857d67461bd7ac4abcc2b371cb7ebc378c [PDF] (accessed on 6 December 2023).

Faircloth, J. B., Capella, L. M., & Alford, B. L., (2015). *The Effect of Brand Attitude and Brand Image on Brand Equity, 9*(3), 61–75. http://Dx.Doi.Org/10.1080/10696679.2001.11501897.

Gamliel, E., & Herstein, R., (2007). The effect of framing on willingness to buy private brands. *Journal of Consumer Marketing, 24*(6), 334–339.

Gangwani, S., Mathur, M., & Shahab, S., (2020a,b). Influence of consumer perceptions of private label brands on store loyalty–evidence from Indian retailing. *Cogent Business and Management, 7*(1), 1751905.

Geçti, F. Y., & Zengin, H., (2013). Examining perceived risk and its influence on attitudes: A study on private label consumers in Turkey. *Asian Social Science, 9*(4), 158–166.

Ghali -Zinoubi, Z., (2021). Effect of organic food perceived values on consumer attitude and behavior in developing Country: Moderating role of price sensitivity. *Pakistan Journal of Agricultural Sciences, 58*(3), 779–788.

Glynn, M. S., & Chen, S., (2009a,b). Consumer-factors moderating private label brand success: Further empirical results. *International Journal of Retail & Distribution Management, 37*(11), 896–914.

Hanifawati, T., Dewanti, V. W., & Saputri, G. D., (2019). The role of social media influencer on brand switching of millenial and gen Z: A study of food-beverage products. *Journal of Applied Management (JAM), 17*(4), 625–638.

Hassan, H., & Rahman, M. S. (2012). *A Conceptual Study of National Brand towards Hypermarket Brand Extension through Brand Asset Valuator, International Conference on Economics, Business Innovation.*

Herstein, R., Tifferet, S., Abrantes, J. L., Lymperopoulos, C., Albayrak, T., & Caber, M., (2012). The effect of personality traits on private brand consumer tendencies: A cross-cultural study of Mediterranean countries. *Cross Cultural Management, 19*(2), 196–214.

Homer, P. M., & Kahle, L. R., (1988). A structural equation test of the value-attitude-behavior hierarchy. *Journal of Personality and Social Psychology, 54*(4), 638–646.

Horvat, S., & Ozretić, D. Đ., (2013). Perceived risk influence on the consumer attitude to private labels in the product's life cycle growth stage. *Economic and Business Review, 15*(4), 267–291.

Iyer, B., (2015). *Big Bazaar: Private Label Push* | Business Standard News. Business Standard. https://www.business-standard.com/article/management/big-bazaar-private-label-push-109120100091_1.html (accessed on 6 December 2023).

Kakkos, N., Trivellas, P., & Sdrolias, L., (2015). Identifying drivers of purchase intention for private label brands. Preliminary evidence from Greek consumers. *Procedia – Social and Behavioral Sciences, 175*, 522–528.

Kumar, R., & Kaushal, S. K., (2017). Examining factors affecting consumers' attitude and purchase intention with special reference to electronic durable goods examining factors affecting consumers' attitude and purchase intention with special reference to electronic durable goods. *NMIMS Management Review, 35*(3), 25–45.

Malaysia, U. K., Kumar, A., Meesala, A., Nair, G., Zainuddin, A., Ridzwan, S. B., Mohd, R., & Ridzwan, S. I., (2018). Assessing the effect of visual merchandising on the relationship between store image and mall choice impact of visual merchandising element son business outcomes of whit e goods and electronics St o… assessing the effect of visual merchandising on the relationship between store image and mall choice. *International Journal of Engineering and Technology (UAE) International Journal of Engineering & Technology*, *7*(4), 231–237.

Measurement and Application of Constructs, (n.d.). Models and Mechanisms of Public Health. Retrieved from https://courses.lumenlearning.com/suny-buffalo-environmentalhealth/chapter/measurement-and-application-of-constructs/ (accessed on 6 December 2023).

Ming-Sung, C. J., Shui-Lien, C. L., Shih-Tse, W. E., & Ying-Chao, L. J., (2007). Do consumers perceive differences among national brands, international private labels, and local private labels? The case of Taiwan. *Journal of Product & Brand Management*, *16*(6), 368–376.

Mishra, S., Malhotra, G., & Saxena, G., (2020). In-store marketing of private labels: Applying cue utilization theory. *International Journal of Retail & Distribution Management*.

Mostafa, R. H. A., & Elseidi, R. I., (2018). Factors affecting consumers' willingness to buy private label brands (PLBs) applied study on hypermarkets. *Spanish Journal of Marketing – ESIC*, *22*(3), 341–361.

Nepul Raj, A., & Jayaprakash, D. (2021). *Meme Usage Culture On Social Issues Among Youth*. *12*(11), 2099–2107.

Olson, J. C., & Jacoby, J., (1972). *Cue Utilization in the Quality Perception Process*. ACR Special Volumes, SV-02.

Oosthuizen, M. D., (2015). The relationship between perceived price and consumers' purchase intentions of private label wine brands. In: *African Journal of Hospitality, Tourism and Leisure* (Vol. 4, No. 2).

Ramadhan, M. D., & Muthohar, M., (2019). *The Influence of Perceived Price, Perceived Quality, Brand Image, and Store Image on the Purchase Intention of Hypermart Private Label*. *308*(Insyma), 140–143.

Redler, J., & Hurth, J., (2020). Gen Z's altruism and performance values link to private label preference: First evidence from a German drugstore sample. *The Business and Management Review*, *11*(1).

Retailing: Critical Concepts. 3,2. Retail Practices and Operations – Google Books, (n.d.). Retrieved from: https://books.google.co.in/books?hl=en&lr=&id=Rt96wAigg2oC&oi=fnd&pg=PA98&dq=martineau+store+image&ots=pTt65DBFzM&sig=9Y4udkjr9a-ZMG-6huFX9UN7-ZM&redir_esc=y#v=onepage&q=martineaustoreimage&f=false (accessed on 6 December 2023).

Riboldazzi, S., Capriello, A., & Martin, D., (2021). Private-label consumer studies: A review and future research agenda. *International Journal of Consumer Studies*, 844–866.

Richardson, P. S., (1997). Are store brands perceived to be just another brand? *Journal of Product & Brand Management*, *6*(6), 388–404.

Richardson, P. S., Dick, A. S., & Jain, A. K., (1994). Extrinsic and intrinsic cue effects on perceptions of store brand quality. *Journal of Marketing*, *58*(4), 28.

Ruiz-Real, J. L., Gázquez-Abad, J. C., Esteban-Millat, I., & Martínez-López, F. J., (2017). The role of consumers' attitudes in estimating consumer response to assortment composition: Evidence from Spain. *International Journal of Retail and Distribution Management*, *45*(7, 8), 782–807.

Sarkar, S., Sharma, D., & Kalro, A. D., (2016). Private label brands in an emerging economy: An exploratory study in India. *International Journal of Retail and Distribution Management*, *44*(2), 203–222.

Shannon, R., & Mandhachitara, R., (2008). Causal path modeling of grocery shopping in hypermarkets. *Journal of Product and Brand Management, 17*(5), 327–340.

Srinivasan, S. S., & till, B. D., (2002). Evaluation of search, experience, and credence attributes: Role of brand name and product trial. *Journal of Product & Brand Management, 11*(7), 417–431.

Thangavel, P., Pathak, P., & Chandra, B., (2021). Millennials and generation Z: A generational cohort analysis of Indian consumers. *Benchmarking, 28*(7), 2157–2177.

The Theory of Planned Behavior. (n.d.). Retrieved from: https://sphweb.bumc.bu.edu/otlt/mph-modules/sb/behavioralchangetheories/BehavioralChangeTheories3.html (accessed on 6 December 2023).

Tifferet, S., & Herstein, R., (2010). The effect of individualism on private brand perception: A cross-cultural investigation. *Journal of Consumer Marketing, 27*(4), 313–323.

Turner, A., (2015). Generation Z: Technology and social interest. *Journal of Individual Psychology, 71*(2), 103–113.

Underwood, R. L., Klein, N. M., Burke, R. R., Kelley, Professor, E. W., & Administration, B., (2001). Packaging communication: Attentional effects of product imagery. *Journal of Product and Brand Management.*

Vahie, A., & Paswan, A., (2006). Private label brand image: Its relationship with store image and national brand. *International Journal of Retail and Distribution Management, 34*(1), 67–84.

Valaskova, K., Kliestikova, J., & Krizanova, A. (n.d.). *Consumer perceptions of private label products: an empirical study. Journal of Competitiveness, 10*(3), 149–163.

Veloutsou, C., Gioulistanis, E., & Moutinho, L., (2004). Own labels choice criteria and perceived characteristics in Greece and Scotland: Factors influencing the willingness to buy. *Journal of Product & Brand Management, 13*(4), 228–241.

Yang, Z., & Peterson, R. T., (2004). Customer perceived value, satisfaction, and loyalty: The role of switching costs. *Psychology & Marketing, 21*(10), 799–822.

PART IV

Green Marketing and the New Consumption Pattern

CHAPTER 13

Exploring Buying Intention and Brand Loyalty of Millennials and Gen Z Toward Sustainable Beauty and Personal Care Products Using a Green Mix Approach

DEBOSHREE BARMAN SEAL, SUDIN BAG, and KOUSIK MANDAL

Department of Business Administration, Vidyasagar University, Midnapore, West Bengal, India

ABSTRACT

The term 'Green' has become increasingly important in purchase decision process for consumers as the preference for eco-friendliness, reduction in the quantity of the waste and toxic consumption, consumption of recycled products is on the top priority in the recent time. This research chapter was undertaken to gauge how Green Marketing Mix is affecting the buyer's intent to purchase BPC products amongst generation Y and Z in the Indian subcontinent. Furthermore, the study also surveys the importance of consumers' grasping prowess of knowledge of Green Brand and how it is shaping their intention of purchase. Moreover, the relationship between purchase intention and brand loyalty was also examined to identify the buying behavior of generation Y and Z with special emphasis on beauty and personal care products. For deriving the empirical results of the aforementioned study, the researcher has used an explanatory research methodology. An online survey through a Google form link with a self-administered structure research questionnaire is shared among the Millennial and Gen-Z to capture their valuable feedback. A total of 230 consumers' responses were used to gauge the effectiveness of the green marketing mix on purchase intention. Data

Sustainability in Marketing Practice: Strategies for Industry 4.0.
Subrata Chattopadhyay, Udit Chawla, (Eds.)

analyzes were done through structural equation modeling by Smart PLS version 3.3.0. The result suggests that the sustainable product or products which are deemed to be Green, related promotional campaigns and activities have positively impacted green buying decision of beauty and personal care products. But green price and green place negative impact found for purchasing green BPC products. However, comprehension and cognition about sustainable green products and attitude towards tenable products positively impact green buying decision. Moreover, the result reveals that; those buyers who show a positive attitude towards Green tenable products also shows strong Brand affinity which, in turn leads to consumer-based brand loyalty. The overall knowledge of the green product lifecycle and how it affects purchase intention which leads to brand loyalty, will help the marketers to properly distinguish potential or prospective consumers and then developing counter made strategies for attracting and engaging the consumer with higher consumer retention. It will help the organizations to portray an Environment conscious member of the society which in turn will favor them in attracting consumers who are ecologically sound and also convert prospective clientele to regular consumer. The ramification of this study will comprehend in understanding different factors affecting the green purchase decision of consumers; how, the overall purchase experience, usability, and green disposability of the product will lead to brand loyalty.

13.1 INTRODUCTION

India, the largest and most populous democracy in the world, has achieved leaps and bounds in the last few decades. Economically, India is in a stable position; its extended growth has steadily increased over the last 50 years without any major economic or financial setbacks. This growth is the result of a strong establishment and stability of key sectors, i.e., agriculture, industry, and services, which have enabled India to attract more foreign investments, increase exports, and also drive the country's development through internal consumption (Gupta & Blum, 2018).

The stable and growing economy has uplifted millions of people out of poverty and improved their lifestyles, propelling India to become a key global economic player. The enactment of a progressive set of fiscal and trade reforms and policies has been a boon for strong economic activities, which have successfully helped India in controlling and disintegrating inflation, as well as defending against fiscal and current account deficits (OECD, 2021).

India approximately boasts one of the largest young populations in the world; millennials account for 34% at 440 million, while Gen Z accounts for 32% at 472 million of the country's total population (Sharma, 2021; Verma, 2020). According to a report published by Morgan Stanley, the rising income of the Indian young generation poses a great opportunity for consumer industries. Millennials and Gen Z, who were born between 1982 AD and 2012 AD, comprise just about one-third of the Indian populace and also fuel the country's growth by accounting for 46% of the workforce. The brimming youth, thanks to the stable economy and society, have emerged as the main financial providers in most households, contributing almost 70% of the total family earnings (Morgan Stanley Research, 2017).

India, though geographically one-third the size of China, is set to become the most populous nation in the world by 2050 (Population Reference Bureau (PRB), 2001). This huge population burst may pose a new set of challenges in the coming time, such as a shortage of necessary resources for societal development, which may lead to a depletion of environmental resources like land erosion, water scarcity, and deforestation (Sreen et al., 2018).

Due to this, in the past decades, with the ease of availability of information, both organizations and consumers have expressed serious disquiet over the environmental repercussions of products. Hence, over the years, we are now experiencing global attention towards environmentally friendly products that either have negligible or zero negative impact on the environment. Products that are speculated to be "Green" or "Environment Friendly," such as energy-efficient consumer durables, organic foods, organic beauty and personal care products, and organic pesticides, are finding more takers (Bhatia & Jain, 2013). Present-day consumers are becoming more aware and thoughtful towards the socio-ethical contemplation of organizations (Chen, 2001).

Indian Marketers gradually understand the importance of Green Marketing (Bhatia & Jain, 2013). Consumers, on the other hand, are now looking up to organizations as an environmentally social player in society rather than just being a profit-making machine. Therefore, organizations are now shifting towards a sustainable business model, focusing on establishing green practices throughout the Product Life Cycle and offering at par or better alternative "Green Product" compared to Traditional product (Sreen et al., 2018). "Green products" are made by implementing energy-efficient manufacturing processes, eliminating the use of toxic and harmful materials, which protect and conserve natural resources (Ottman, 1992; do Paco & Raposo, 2009). It is evident that ecologically sound buyers or consumers are more likely or susceptible to purchase pollution-free products (Sarumathi, 2014), referring to this phenomenon as Green Consumption (Anderson & Cunningham, 1972). The underlying concept behind this new phenomenon is that while purchasing price and quality are not the only driving factors (Monroe, 1976), norms, values, and beliefs also play a crucial role in the purchase decision (Caruana, 2007).

As a result, the traditional 4Ps of Marketing (Product, Price, Place, and Promotion) prescribed by Kotler (1967) widely used in marketing research (for instance, Bag & Biswas, 2013) are modified, keeping environmental factors in mind, giving birth to Green Marketing Mix.

Green Marketing, though in its nascent stage, is an extensive subject, which has a broad connotation ranging from trade policies to public policies. In general, consumers associate terms like Upcycled, Recyclable, Paraben Free, Phosphate Free, and Eco-Friendly with Green Marketing. Triumphant Green Marketers and Green-Preneurs (Green Entrepreneurs) approach their consumers as responsible green members of society rather than traditional material goods consumers, in turn creating more value for consumers throughout the product life cycle. The five main advantageous factors of using green products are:

- Manufacturing efficiency leading to cost effectiveness;
- Health and safety;
- Performance;
- Hedonic status and prestige symbol;
- Convenience.

Due to which green marketing is now shifting from niche marketing to mass marketing (Pandey & Singh, 2012).

From the previous research work undertaken by various scholars, it has come forward that there have been very few and far between works which have been done to understand the "buying intention" of "green products" in the context of the Indian market; due to which there are only a handful of literature works available in this sector. Research work shows that there is a lacuna in comprehending and creating a perception of Green Products in India (Seal and Bag, 2022). Compared to the awareness level of consumers of developed countries like the USA (73%), Canada (78%); 63% of Indian buyers are well informed of Sustainable Products and their implications. Awareness development of consumers can resolve the obstacle by formulating better Green Products (Bonini & Oppenheim, 2008) as more information will help to build a positive and informative outlook towards green products (Pandey & Singh, 2012).

The good news is the Indian Government has over the years implemented schemes like the Ecomark scheme in 1991 (Bahadur & Manjunath, 2019), environment protection and awareness acts, and have educated the people about the ill effects of pollution (Jafri, 2020). Worldwide, most of the nations are having serious concern over the detrimental adversarial ill-effects of human engagement on the environment around us; hence, they have started to implement different Green initiatives to safeguard the environment (McMichael, 2013). Upon the comprehension and realization of the organization's harmful act on the society and environment, the concept of Sustainable Development (Pisani, 2006; Waas et al., 2011) has emerged to focus on development which is beneficial and also has very little effect on the environment. By incorporating Sustainability, it boosts the well-being and mindset of the individual and creates a positive state of awareness and prosperity for the nation as a whole; this idea has given birth to the new societal concept of "Going Green" (Sonigo et al., 2012).

The upshot of this trend has led to the advent of "Green Products." The term "Green Product" means the products which are produced in an environmentally friendly manufacturing process by minimizing or completely nullifying the usage and ill effects of toxic materials and ecologically harmful products and processes; thus, tenable products protect the environment and promote a healthy lifestyle. Therefore, ecologically aware buyers have shown an affirmative attitude and affinity with respect to "Green Products." Emerging new consumer behavior, awareness towards products, government policy, influence of environment warriors and NGOs have prompted the organizations to gradually develop 'Sustainability' and produce "nonpolluting" or "Green Products" (Bhardwaj et al., 2020).

In the beauty and personal care sector also, consumers and organizations are warming up to the idea of alternative or natural cosmetics compared to the age-old traditional environmentally harmful products (Amberg & Fogarassy, 2019). The focus of this empirical research study is to comprehend how the new concept of Green Marketing Mix is getting adopted by organizations and its effect on the purchase behavior of Gen Y and Gen Z consumers of India (Smith & Brower, 2012), and how this newfound affinity of green mannerism will affect the brand loyalty in the field of Beauty and Personal Care Industry.

13.2 LITERATURE REVIEW AND HYPOTHESIS GENERATION

The marketing process of environmentally safe products is known as "green marketing;" green, eco-friendly are the new perspective which tries to bridge the gap between conventional marketing approach and the ecological-societal dimensions of Green Marketing (Belz & Peattie, 2009). In their research chapter (Vaibhav et al., 2015) have infused the eco-friendly concept with the traditional marketing mix thus green marketing mix model contains 4Ps:

- Green product;
- Green price;
- Green place; and
- Green promotion.

13.2.1 EFFECT OF GREEN PRODUCT ON PURCHASE DECISION

Organizations should produce such products for mass consumer base which are not only beneficial to the environment but even after its life cycle it could be recycled or upcycled for the betterment of the environment. Organizations are nowadays incorporating upgraded machineries and principles to eliminate the detrimental aspects of manufacturing on the environment. The Green product design not only be customer friendly it also needs to be ecofriendly and non-vulnerable; the products also need to be at par or beyond the value provided by the conventional products (Vaibhav et al., 2015). As Green Products are gradually becoming the new norm it is important to address the gap between the purchase intention and real purchases of them (Auger & Devinney, 2007; Carrington et al., 2014). This study will focus on the purchaser's behavior regarding the perception of tenable product (Brough et al., 2016; Lin & Chang, 2012) and its effect on final purchase intention.

- **H1:** Green products positively affects the purchase intention of consumers.

13.2.2 EFFECT OF GREEN PRICE ON PURCHASE DECISION

Tenable eco-products are extravagant than the usual run-of-the-mill products; but if the buyer or user who are well versed about the added extra value and benefits of tenable products, they are more than happy to shell out extra cash for availing the product (Abzari et al., 2013). The price band difference between conventional products and green products is refereed as Green Price

(Davari & Strutton, 2014). Due to high procurement cost of sustainable raw materials the final product becomes costlier (Afzan, 2016; Hwang et al., 2017) the onus is on the organizations to educate the consumers about the benefits of using Green Products and make them well informed about the special feature and value addition of these products. Hence, until cost of the production comes down the goal of the organizations should be to attract more consumers who prefer to use green products rather than targeting the broader spectrum of targeting price sensitive consumers (Meng et al., 2021). This research chapter will find out if green price at all affects the intention to purchase or there is no effect at all.

- **H2:** Green price positively affects the purchase intention of consumers.

13.2.3 EFFECT OF GREEN PLACE ON PURCHASE DECISION

Green Place enacts a pivotal aspect in the Green Marketing Mix. It involves a complex set of distribution logistics which starts from the raw material procurement to effective handling of distribution network to the final purchase or consumption point minimizing or eliminating the carbon foot-print. This method involves various stages like:

- Procurement of sustainable products through green supply logistics for example use of EV (electronic vehicles) for transportation of goods.
- Implementation of green manufacturing process and technology; enabling optimal use of raw material, proper utilization of energy and effective waste management.
- Creation of green sales network that is implementing optimal beat map planning of routes.
- Effectively designing and implementing in a cost-effective ecofriendly manner to gather information and returned products from consumption point to manufacturing point (Zhu et al., 2008).

Therefore, green place effectively refers to green distribution logistics; hence producers must ensure that green products are available for conscious consumers at green locations (Sohail, 2017; Mishra & Sharma, 2012). This study chapter will focus on finding a co-relation between green place and consumer purchase intention.

- **H3:** Green place positively affects the purchase intention of consumers.

13.2.4 EFFECT OF GREEN PROMOTION ON BUYING DECISION

Refers to the communication between the organization and consumers where organizations provide their ecofriendly initiatives and benefits of their green products to consumers (Polonsky & Rosenberger, 2001); consumers on the other hand assess these promotional activities and try to comprehend about its effect on the environment (Kinoti, 2011). Green promotion focuses on the promotional activities of Sustainable polices and standards adopted by the organization; like the organization should highlight if they possess a CP certificate or if it is ISO 14000 certified (Adhikari et al., 2016). The focus of this study will be to establish an association between Tenable Promotion and Determination to Purchase.

- **H4:** Green promotion positively affects the purchase intention of consumers.

13.2.5 EFFECT OF BRAND KNOWLEDGE ON PURCHASE INTENTION

Brand Knowledge plays a pivotal part in shaping consumer's mindset about a particular product. Consumer tends to reciprocate with brands and products as a whole when they associate themselves positively with the message during promotional campaigns. Based on these ad campaigns brands create successful positioning of their product in consumer's mind which results in positive purchase behavior in present or future (Asadollahi et al., 2011). According to research done by Kapferer (2004) based on how a brand is portraying its benefits and value through promotional campaigns the brand subconsciously develops an image in consumer's mind. Hence, agenda of this study will be to identify whether the same phenomenon is observed in India while consumer purchase green BPC products or a different scenario comes to the picture.

- **H5:** Brand knowledge positively influences purchase decision.

13.2.6 INFLUENCE OF ATTITUDE TOWARDS SUSTAINABLE PRODUCT ON PURCHASE DECISION

The theory of planned behavior (TPB) expresses that determination to purchase is the main driving factor behind the final purchase behavior of consumers (Phau & Teah, 2009). Attitude is a socially learned behavior; which propels the individuals to act in a positive or negative manner upon a particular product or scenario In the context of Marketing; Organizations and Marketers need to gauge the consumers attitude by observing their previous buying pattern, use of test marketing as attitude is a societal concept which cannot be measured physically (Huang et al., 2004). The research study will try to evaluate if Attitude towards Green BPC products affects the Purchase intention of consumers of India.

- **H6:** Attitude towards product positively influences purchase decision.

13.2.7 PURCHASE INTENTION AFFECTS BRAND LOYALTY

Brand loyalty refers to the consumer behavior where out of all competitive brands consumer repeatedly purchases products of a same brand over a time and may also acts as a brand advocate and promoter (Aaker, 1991). From the view point of the organizations brand loyalty plays a crucial role as it helps in developing entry barriers for other potential players of the market, enables a competitive edge over others and increasing sales number, market share (Ballester & Munuera-Aleman, 2001) and lowering the price sensitive barrier of consumers (Rowley, 2005). Hence, aim of this research work is to establish an association between the two metrices.

- **H7:** Purchase intention affects brand loyalty.

13.2.8 CONCEPTUAL FRAMEWORK (FIGURE 13.1)

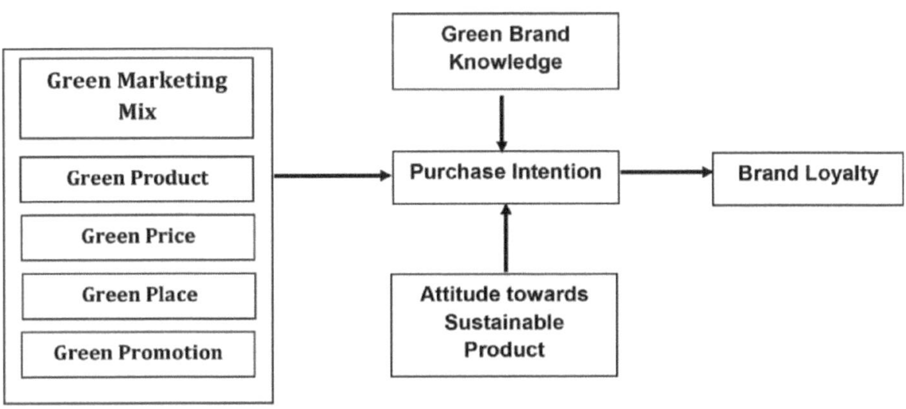

FIGURE 13.1 Conceptual model.

13.3 RESEARCH METHODOLOGY

13.3.1 RESEARCH DESIGN

By studying upon prior research and literary works hypothesis have been generated through the support of fact-finding data and relevant theories. For this study we have employed Explanatory Research Design; which will help to diagnose and establish proper relationship between different factors. To have a proper efficient representation of the population in an economic manner the researcher has deployed a survey research method for this study.

13.3.2 DATA COLLECTION AND PROCEDURE

For the current study both Primary and Secondary data were considered. Various literary works, books, journals, and study paper have acted as source of secondary data. With the help of secondary data, past theories and literary works have been investigated to formulate the Literature Review. Survey procedure was employed to gather Primary Data from participants. Data accumulation process has been done by researchers to authenticate the purpose of this research and objectives.

Researchers have sent a copy of questionnaire to participants for filling up the same and send back to the researchers through the use of Google Forms. The questionnaire was based on the prior purchase experience of Green Products by the consumers and to gauge their response 5-point Likert Scale system have been deployed ranging from:

- Strongly agree (Representation value: 5);
- Agree (Representation value: 4);
- Neutral (Representation value: 3);
- Disagree (Representation value: 2);
- Strongly disagree (Representation value: 1).

The questionnaire distribution and data collection were done in the month of February 2022; due to COVID restrictions face to face interview was avoided.

13.3.3 DATA SAMPLING

For authentication and verification of this research both male and female participants were chosen from Generation Y and Z of India. This particular generation was of interest as compare to other this particular set of people are more knowledgeable about the intricacies of technology (Immordino-Yang, Christodoulou, & Singh, 2012; Martin, 2005) also due to higher disposable income they are more demanding and shows strong affination towards brand of their choice (Straus, Howe, & Markiewicz, 2006; Morton, 2002). The age group of this study sample ranged from Age 18 years to 42 years old. A total of 230 participants were considered for this research work; self-selection sampling method was used for data collection. Interested participants were sent a hyperlink of Google form through e-mail, WhatsApp, Facebook, and Instagram.

13.3.4 DATA QUESTIONNAIRE

The questionnaire for this empirical research study was designed in a self-administered structured research method; all the questions asked was close ended. The modules used in this research have been derived from previous research study by Munamba & Nuangjamnong (2021). Partial least square equation modeling (PLS-SEM) (Bag, Aich, & Islam, 2020) technique was used as the principal statistical modeling tool.

13.4 DATA ANALYSIS

13.4.1 EVALUATION OF RELIABILITY AND VALIDITY

Convergent validity, discriminant validity and reliability of measure have been considered for establishing structural measurement model. For analyzing the reliability of the variables, the researchers have considered Cronbach's Alpha and composite reliability (CR) for the study. Table 13.1 representing the reliability and validity of variables, shows that, the value of Cronbach's alpha and CR of all the variables are more than the proposed value of 0.7 and 0.5 (Bag & Omrane, 2021, 2022). The outcome of reliability and validity table also depict that the average variance extracted (AVE) of all the variance are greater than 0.5, which is the threshold value (Hair et al., 2017; Sinha & Bag, 2022). The analysis of Table 13.1 represents that the reliability and validity of the variables are well established and are accepted.

TABLE 13.1 Reliability and Validity

	Cronbach's Alpha	Composite Reliability	Average Variance Extracted (AVE)
AT	0.759	0.851	0.595
BL	0.724	0.828	0.547
GBK	0.824	0.885	0.659
GPL	0.789	0.876	0.702
GPM	0.766	0.851	0.589
GRI	0.778	0.870	0.692
GRP	0.794	0.846	0.587
PI	0.818	0.874	0.583

Note: AT: Attitude towards green product; BL: Brand loyalty; GBK: Green brand knowledge; GPL: Green place; GPM: Green promotion; GRI: Green price; GRP: Green product; and PI: Purchase intention.

Discriminant validity in a research study is developed in order to find out the distinctiveness of the variables. It demonstrates that the variables of the conceptual framework have their distinctive identity and the variables are not highly correlated with each other. The researchers have used two different techniques of discriminant validity such as Fornell-Larcker criterion and heterotrait-monotrait (HTMT) Ratio for the research study. The Fornell-Larcker Criterion is mainly employed to analyze the degree of shared variance amongst the latent variables of the study (Fornell & Larcker, 1981; Bag et al., 2021). From Table 13.2, it is evident that all the diagonal values of the variables are greater than all the values in the similar row and column. This portrays that all the variables are distinct in nature and are not related to each other.

TABLE 13.2 Discriminant Validity (1) Fornell-Larcker Criterion

	AT	BL	GBK	GPL	GPM	GRI	GRP	PI
AT	0.771	–	–	–	–	–	–	–
BL	0.650	0.740	–	–	–	–	–	–
GBK	0.763	0.578	0.812	–	–	–	–	–
GPL	0.626	0.511	0.600	0.838	–	–	–	–
GPM	0.605	0.630	0.501	0.615	0.767	–	–	–
GRI	0.769	0.705	0.791	0.680	0.569	0.832	–	–
GRP	0.574	0.729	0.531	0.652	0.707	0.646	0.766	
PI	0.610	0.739	0.589	0.496	0.557	0.540	0.593	0.764

The researchers have analyzed the HTMT ratio, which evaluates the correlation between two dormant variables. HTMT helps to find out the discriminant validity of the variables of the model. As per Henseler et al. (2015); Ghorai et al. (2021), the value of HTMT of each dimension should be less than 0.9. Table 13.3 represents that the HTMT ration of each dimension are lower than 0.9. Hence, through Tables 13.2 and 13.3 it is evident that the discriminant validity of all the variables is accepted.

TABLE 13.3 Discriminant Validity (2) HTMT Ratio

	AT	BL	GBK	GPL	GPM	GRI	GRP	PI
AT	–	–	–	–	–	–	–	–
BL	0.872	–	–	–	–	–	–	–
GBK	0.854	0.740	–	–	–	–	–	–
GPL	0.827	0.679	0.729	–	–	–	–	–
GPM	0.804	0.844	0.619	0.776	–	–	–	–
GRI	0.864	0.830	0.872	0.875	0.719	–	–	–
GRP	0.787	0.807	0.646	0.764	0.879	0.809	–	–
PI	0.761	0.845	0.713	0.596	0.681	0.650	0.625	–

13.4.2 TESTING OF HYPOTHESIS

Succeeding the evaluation of the outer model, the hypothesis testing was performed. Figure 13.2 and Table 13.4 depicts the association amongst the different variables. The consequence of the analysis has portrayed that Green Product, Green Promotion, Green Brand Knowledge and Attitude towards Sustainable Product have positively relation between them which eventually is influencing the Purchase Decision of Green Products. The results also show that purchase decision has positive association influencing Brand Loyalty of the customers for Green Product.

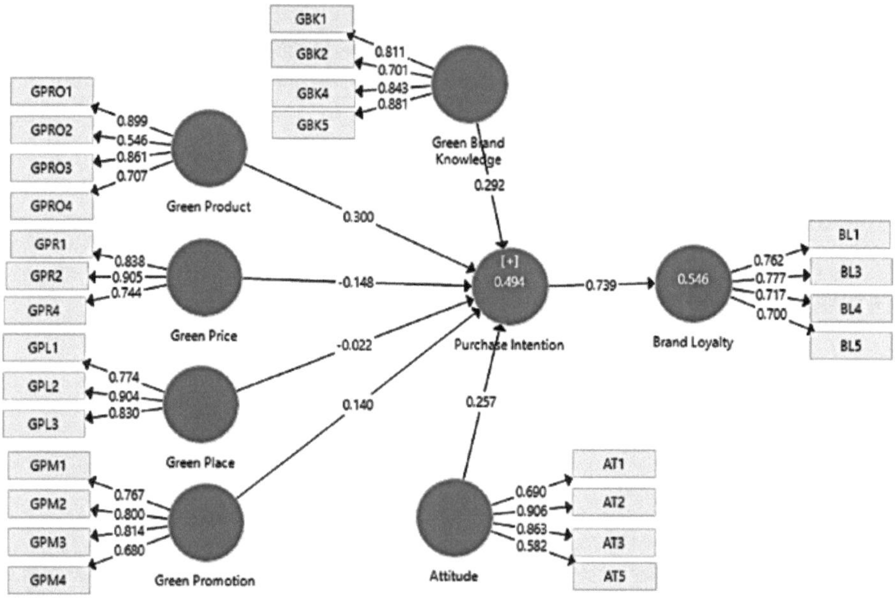

FIGURE 13.2 Path analysis model.

TABLE 13.4 Result of Path Analysis

Hypothesis Path	Coefficient of β (P-Value)	Remarks
Green product → Purchase intention	0.300 (0.05)	Supported
Green price → Purchase intention	–0.148 (0.05)	Negatively support
Green place → Purchase intention	–0.022 (0.05)	Negatively support
Green promotion → Purchase intention	0.140 (0.05)	Supported
Green brand knowledge → Purchase intention	0.292 (0.05)	Supported
Attitude → Purchase intention	0.257 (0.05)	Supported
Purchase intention → Brand loyalty	0.739 (0.05)	Supported

13.5 CONCLUSION AND DISCUSSION

The result suggests that the Green Product and Green Promotion have positively impacted buyer's decision to buy tenable products of beauty and personal care industry.

Green Products; based on the results, it has the highest relationship with green purchase intention in comparison to other factors in Indian BPC sector. This result has furthermore solidified the previous findings of research work of Mahmoud et al. (2017); Hossain & Khan (2018) as well as Mahmoud (2018), who also emphasized amongst all other metrices of Green Marketing Mix, "Green Products," has the highest correlation with Green Purchase.

Green Promotion; based on the results, have second highest correlation with Green Purchase Decision in Indian BPC sector. It corroborates the previous findings of Mahmoud (2018); Mahmoud et al. (2017); Hossain & Khan (2018); and Ansar (2013) which states that Green Promotion has a ginormous effect on the consumer intent to purchase.

But outcome of this study, has revealed that Green Price and Green Place have an adverse ramification on the consumer Green Purchase intention in the context of Indian BPC industry.

Green price; based on the results, negatively correlates with Green Purchase Decision. This finding also indicates to the same scenario discovered by previous researchers (Aertsens et al., 2011) which states that even ecologically sound consumers are finnicky about product price which acts as a hindrance for their intention to purchase the Green alternative products.

Green place; based on the results, negatively ramifies consumer's Green Purchase Decision in the context of Indian BPC industry. These findings align with the result of past research (Barbarossa & Pastore, 2015), that though consumers show strong affinity towards green products but due to lack of proper green distribution network infrastructure and availability of products in an eco-friendly manner they refrain from purchasing the same.

Brand knowledge and attitude positively affluences the consumer's intention to purchase Green BPC products in Indian Market. This finding furthermore proves the fact that more knowledgeable the consumer becomes about the benefits of using Green Products the propensity to purchase increases (Suki, 2016), Green Brand knowledge is one of the key factors of driving the purchase intention. Furthermore, it also confirms the findings of Huang et al. (2014) consumers who have a strong attitude towards green products purchases the eco-friendly products more frequently.

Moreover, the result shows that green purchase intention positively influences green consumer-based brand loyalty in Indian BPC sector; which aligns perfectly with the conclusion of past research work (Fazeen et al., 2020; Lin et al., 2017). This solidifies the fact that more the consumers purchase Green Products they become satisfied with the value proposition and develops a strong bond with the Brand.

13.6 RESEARCH IMPLICATION

This research was undertaken to evaluate the association between Green Marketing Mix and Green Purchase Intention and how it affects the Consumers Brand Loyalty in the context of Indian BPC sector. From the findings of this research, we can conclude the following:

- Organizations must ensure that Green Products must be developed by procuring sustainable raw materials and manufactured in a sustainable eco-friendly manner; also ensuring throughout the product life cycle it never have a harmful effect on its surroundings (Solaiman et al., 2015).
- Green promotion also plays a pivotal role in shaping the consumer mindset about the value addition provided by the Green Products. Promotional activities such as BTL, ATL, TTL activities (Mahmoud, 2018) must be done in an eco-friendly sustainable manner such as eliminating the use of paper or toxic materials in their campaigns; use green materials for making Flyers, Pamphlets, and other promotional materials.
- Pricing of green products, as they are priced higher than conventional products may sometimes hinder the purchase decision. Hence, it is advised to implement energy efficient manufacturing process to lower the production cost and offer the product at a competitive price; this will help in attracting a broader range of first-time buyers.
- The research findings highlight that consumer are presently not happy about how the Green Place vertical is managed. Therefore, it is suggested to promote Green Products in places which maintains the standards of green environment; also, the sales team needs to be trained to educate the consumers about Green Product benefits and its usage mechanism. Organizations also need to develop green end to end supply chain logistics to furthermore reduce the Carbon footprint.

The overall knowledge of the green product lifecycle and how it affects purchase intention which leads to brand loyalty, will help the marketers, to distinguish between traditional and ecologically conscious consumers and accordingly develop strategies for attracting and engaging the consumer with higher consumer retention. It will help the organizations to curate an image of socially conscious member of the society.

KEYWORDS

- **brand loyalty**
- **data collection**
- **data questionnaire**
- **data sampling**
- **green brand knowledge**
- **green marketing mix**
- **sustainable product**

REFERENCES

Aaker, D. A., (1991). *Managing Brand Equity*. The Free Press.

Abzari, M., Safari, F., Sharbiyani, A. A., & Morad, A. P., (2013). Studying the effect of green marketing mix on market share increase. *Advances in Environmental Biology, 7*(10), 2981–2991.

Adhikari, A., Biswas, I., & Avittathur, B., (2016). Green retailing: A new paradigm in supply chain management. *Handbook of Research on Strategic Supply Chain Management in the Retail Industry*. IGI Global.

Aertsens, J., Mondelaers, K., Verbeke, W., Buysse, J., & Van, H. G., (2011). The influence of subjective and objective knowledge on attitude, motivations, and consumption of organic food. *British Food Journal, 113*(11), 1353–1378.

Afzan, A. Z., (2016). *A Model for Implementation of Green Construction*. PhD Dissertation, University Technology Mara, Shah Alam, Malaysia.

Amberg, N., & Fogarassy, C., (2019). Resources communication green consumer behavior in the cosmetics market. *Resources, 8*(3). 10.3390/resources8030137.

Anderson, W. T., & Cunningham, W. H., (1972). The socially conscious consumer. *Journal of Marketing, 36*(3), 23–31.

Ansar, N., (2013). Impact of green marketing on consumer purchase intention. *Mediterranean Journal of Social Sciences, 4*(11), 650–655. https://doi.org/10.5901/mjss.2013.v4n11p650.

Asadollahi, A., & Hanzaee, K. H., (2011). Investigating the effect of brand knowledge and brand relationships on purchase behavior of customers. *World Applied Sciences Journal, 13*(9), 2012–2020.

Auger, P., & Devinney, T. M., (2007). Do what consumers say matter? The misalignment of preferences with unconstrained ethical intentions. *Journal of Business Ethics, 76*, 361–383.

Bag, S., & Biswas, D., (2013). Four P's compass model in new paradigm of marketing mix. *International Journal of Applied Services Marketing Perspectives, 2*(1), 273–282.

Bag, S., & Omrane, A., (2021). The relationship between the personality traits of entrepreneurs and their decision-making process: The role of manufacturing SMEs' institutional environment in India. In: *Forum Scientiae Oeconomia* (Vol. 9, No. 3, pp. 103–122).

Bag, S., & Omrane, A., (2022). Corporate social responsibility and its overall effects on financial performance: Empirical evidence from Indian companies. *Journal of African Business, 23*(1), 264–280.

Bag, S., Aich, P., & Islam, M. A., (2020). Behavioral intention of digital natives" toward adapting the online education system in higher education. *Journal of Applied Research in Higher Education, 14*(1), 16–40.

Bahadur, B., & Manjunath, S., (2019). Green marketing in India: An overview. *Restaurant Business, 118*(8), 160–171.

Ballester, E., & Munuera-Alemán, J. L., (2001). Brand trust in the context of consumer loyalty. *European Journal of Marketing, 35*(11, 12), 1238–1258.

Barbarossa, C., & Pastore, A., (2015). Why ecologically sound consumers do not purchase green products: A cognitive mapping approach. *Qualitative Market Research, 18*(2), 188–209.

Belz, F., & Peattie, K., (2009). *Sustainability Marketing: A Global Perspective*. John Wiley & Sons.

Bhardwaj, A., Garg, A., Ram, S., Gajpal, Y., & Zheng, C., (2020). Research trends in green product for environment: A bibliometric perspective. *International Journal of Environmental Research and Public Health, 8469*(17), 1–21.

Bhatia, M., & Jain, A., (2013). Green marketing: A study of consumer perception and preferences in India. *Electronic Green Journal, 1*(36).

Bonini, S., & Oppenheim, J., (2008). Cultivating the green consumer. *Stanford Social Innovation Review, 6*, 56–61.

Brough, A. R., Wilkie, J., Isaac, M. S., & Ma, J., (2016). Is eco-friendly unmanly? The green-feminine stereotype and its effect on sustainable consumption. *Journal of Consumer Research, 43*(4), 567–582.

Carrington, M. J., Neville, B. A., & Whitwell, G. J., (2014). Lost in translation: Exploring the ethical consumer intention-behavior gap. *Journal of Business Research, 67*, 2759–2767.

Caruana, R., (2007). A sociological perspective of consumption morality. *Journal of Consumer Behavior, 6*(5), 287–304.

Chen, C., (2001). Design for the environment: A quality-based model for green product development. *Management Science, 47*(2), 250–263.

Davari, A., & Strutton, H., (2014). Marketing mix strategies for closing the gap between green consumers' pro-environmental beliefs and behaviors. *Journal of Strategic Marketing, 22*(7). 10.1080/0965254X.2014.914059.

Do Paço, A., & Raposo, M., (2009). Green' segmentation: An application to the Portuguese consumer market. *Marketing Intelligence and Planning, 27*(3), 364–379.

Economic forecast summary, (2021). *OECD Economic Outlook* (No. 2). Retrieved from: https://www.oecd.org/economy/india-economic-snapshot/ (accessed on 6 December 2023).

Fazeen, A. K., Aubin, D., & Mundol, A., (2020). Factors influencing green brand loyalty and green purchase behavior. *Seventeenth AIMS International Conference on Management*. IIM Kozhikode.

Fornell, C., & Larcker, D. F., (1981). Evaluating structural equation models with unobservable variables and measurement error. *Journal of Marketing Research, 18*(1), 39–50.

Ghorai, S., Sinha, A., & Bag, S., (2021). Exploring the relationship between attitude and purchase intention towards organic food-does 'willingness to pay' mediate the effect? *IITM Journal of Management and IT, 12*(2), 48–58.

Gupta, P., & Blum, F., (2018). *India's Remarkably Robust and Resilient Growth Story.* Retrieved from: https://blogs.worldbank.org/endpovertyinsouthasia/india-s-remarkably-robust-and-resilient-growth-story (accessed on 6 December 2023).

Hair, J. F., Sarstedt, M., Ringle, C. M., & Gudergan, S., (2017). *Advanced Issues in Partial Least Squares Structural Equation Modeling.* Sage. ISBN: 9781483377391.

Henseler, J., Ringle, C. M., & Sarstedt, M., (2015). A new criterion for assessing discriminant validity in variance-based structural equation modeling. *Journal of the Academy of Marketing Science, 43*(1), 115–135. https://doi.org/10.1007/s11747-014-0403-8.

Hossain, A., & Khan, M. Y. H., (2018). Green marketing mix effect on consumers buying decisions in Bangladesh. *Marketing and Management of Innovations, 4*(4), 298–306.

Huang, J., Lee, B. C. Y., & Hsun, H. S., (2004). Consumer attitude toward gray market goods. *International Marketing Review, 21*(6), 598–614.

Huang, Y. C., Yang, M., & Wang, Y. C., (2014). Effects of green brand on green purchase intention. *Marketing Intelligence & Planning, 32*(3), 250–268. https://doi.org/10.1108/MIP-10-2012-0105.

Hwang, B. G., Zhu, L., Wang, Y., & Cheong, X., (2017). green building construction projects in Singapore: Cost premiums and cost performance. *Project Management Journal, 48*(4), 67–79.

Immordino-Yang, M., Christodoulou, J., & Singh, V., (2012). Rest is not idleness: Implications of the brain's default mode for human development and education. *Perspectives on Psychological Science, 7*(4), 352–364. 10.1177/1745691612447308.

Jafri, S., (2020). Indian Government's sustainable initiatives and the millennials behavior. *Journal of Sustainable Development, 13*(1), 78.

Kapferer, J. N., (2008). *The New Strategic Brand Management: Creating and Sustaining Brand Equity Long Term.* Kogan Page Publishers.

Kinoti, M. W., (2011). Green marketing intervention strategies and sustainable development: A conceptual paper. *International Journal of Business and Social Science, 2*, 263–273.

Kotler, P., (1967). *Marketing Management: Analysis, Planning, and Control.* Prentice-Hall, Upper Saddle River.

Lin, J., Lobo, A., & Leckie, C., (2017). Green brand benefits and their influence on brand loyalty. *Marketing Intelligence & Planning, 35*(3), 425–440. 10.1108/MIP-09-2016-0174.

Lin, Y. C., & Chang, C. A., (2012). Double standard: The role of environmental consciousness in green product usage. *Journal of Marketing, 76*(5), 125–134.

Mahmoud, T. O., (2018). Impact of green marketing mix on purchase intention. *International Journal of Advanced and Applied Sciences, 5*(2), 127–135.

Mahmoud, T. O., Ibrahim, S. B., Ali, A. H., & Bleady, A., (2017). The influence of green marketing mix on purchase intention: The mediation role of environmental knowledge. *International Journal of Scientific & Engineering Research, 8*(9), 1040–1048. https://doi.org/10.14299/ijser.2017.09.006.

Martin, C., (2005). From high maintenance to high productivity: What managers need to know about generation Y. *Industrial and Commercial Training, 37*(1). 39–44. 10.1108/00197850510699965.

McMichael, (2013). Globalization, climate change, and human health – NCBI. *N. Engl. J. Med., 368*(14), 1335–1343. https://doi.org/10.1056/NEJMra1109341.

Meng, Z., Zhao, N., Shen, B., & Zhai, C., (2021). Optimal pricing strategy for green products under salience theory. *Economic Research,* 1–24. 10.1080/1331677X.2021.1985572.

Mishra, P., & Sharma, P., (2012). 'Green marketing: Challenges and opportunities for business. *Journal of Marketing & Communication, 8*(1), 35–41.

Monroe, K. B., (1976). The influence of price differences and brand familiarity on brand preferences. *Journal of Consumer Research, 3*(1), 42–49.

Morgan Stanley Research, (2017). India's millennials to drive growth in four key sectors. *Millennials to Anchor India's Growth Story*. Retrieved from: https://www.morganstanley.com/ideas/India-millennials-growth-sectors (accessed on 6 December 2023).

Omrane, A., & Bag, S., (2022). Determinants of customer buying intention towards residential property in Kolkata (India): An exploratory study using PLS-SEM approach, *International Journal of Business Innovation and Research, 28*(1), 119–140.

Ottman, J. A., (1992). Industry's response to green consumerism. *Journal of Business Strategy, 13*, 3–7.

Pandey, K. K., & Singh, P. B., (2012). Green marketing: Policies and practices for sustainable development. *Journal of Management, 5*(1), 22–30.

Phau, I., & Teah, M., (2009). Devil wears (counterfeit) Prada: A study of antecedents and outcomes of attitudes towards counterfeits of luxury brands. *Journal of Consumer Marketing, 26*(1), 15–27. https://doi.org/10.1108/07363760910927019.

Pisani, J. A. D., (2006). Sustainable development historical roots of the concept. *Environmental Science, 3*(2), 83–96.

Polonsky, M. J., & Rosenberger, P. J., (2001). Reevaluating green marketing: A strategic approach. *Business Horizons, 44*(5), 21–30.

Population Reference Bureau (PRB), (2001). *World Population Data Sheet*. Washington, D.C.

Rowley, J., (2005). The four Cs of customer loyalty. *Marketing Intelligence & Planning, 23*(6). 10.1108/02634500510624138.

Sarumathi, S., (2014). Green purchase behavior–a conceptual framework of socially conscious consumer behavior. *Global Journal of Finance Management, 6*(8), 777–782.

Seal, D. B., & Bag, S. (2022). Constituents affecting brand loyalty of sustainable beauty and personal care products. *Marketing and Management of Innovations, 13*(3), 66–77.

Sharma, P., (2021). *The Rise of Indian Millennial*. The Times of India. Retrieved from: https://timesofindia.indiatimes.com/blogs/voices/the-rise-of-the-indian-millennial/ (accessed on 6 December 2023).

Sinha, A., & Bag, S., (2022). Intention of postgraduate students towards the online education system: Application of extended technology acceptance model. *Journal of Applied Research in Higher Education*. https://doi.org/10.1108/JARHE-06-2021-0233.

Smith, K., & Brower, T., (2012). Longitudinal study of green marketing strategies that influence millennials. *Journal of Strategic Marketing, 20*(6), 535–551. 10.1080/0965254X.2012.711345.

Social Innovation Review, (1991). The good news is Indian Government have over the years have implemented schemes like eco-mark scheme.

Sohail, M. S., (2017). Green marketing strategies: How do they influence consumer-based brand equity? *Journal for Global Business Advancement, 10*(3), 229. 10.1504/JGBA.2017.10005507.

Solaiman, M., Osman, A., & Halim, M. S. B. A., (2015). Green marketing: A marketing mix point of view. *International Journal of Business and Technopreneurship, 5*(1), 87–98.

Sonigo, et al., (2012). *Policies to Encourage Sustainable Consumption.* Final report prepared by BIO Intelligence Service for European Commission (DG ENV). Available at: http://ec.europa.eu/environment/eussd/pdf/report_22082012.pdf (accessed on 6 December 2023).

Sreen, N., Purbey, S., & Sadarangani, P., (2018). Impact of culture, behavior, and gender on green purchase intention. *Journal of Retailing and Consumer Service, 4*, 177–189.

Suki, N. M., (2016). Green product purchase intention: Impact of green brands, attitude, and knowledge. *British Food Journal, 118*(12), 2893–2910. https://doi.org/DOI 10.1108/BFJ-06-2016-0295.

Vaibhav, R., Bhalerao, V., & Deshmukh, A., (2015). Green marketing: Greening the 4 Ps of marketing. *International Journal of Knowledge and Research in Management and E-Commerce, 5*, 5–8.

Verma, S., (2020). *It's Time, Gen Z is Ready to Rule the World.* The Indian Express. Retrieved from: https://www.newindianexpress.com/magazine/2020/apr/12/its-time-gen-z-is-ready-to-rule-the-world-2128044.html (accessed on 6 December 2023).

Waas, T., Huge, J., Verbruggen, A., & Wright, T., (2011). Sustainable development. A birds eye view. *Sustainability, 3*(10), 1637–1661.

Zhu, Q., Sarkis, J., & Lai, K., (2008). Confirmation of a measurement model for green supply chain management practices implementation. *International Journal of Production Economics, 111*(2), 261–273. 10.1016/j.ijpe.2006.11.029.

CHAPTER 14

Ecotourism: A Study of the Relationship Between Demographic Factors and Satisfaction Among Customers Toward Ecotourism Using the Chi-Square Application

AMAAN SINGH,[1] SHUBHAM KUMAR SHAW,[1] SAHIL SHAW,[1] PRIYANK KUMAR AGARWAL,[1] and AYUS MEHTA[2]

[1]*Techno India University, Kolkata, West Bengal, India*

[2]*The Calcutta Anglo Gujarati College, Kolkata, West Bengal, India*

ABSTRACT

Tourism might be a social, cultural, and financial improvement that involves the movement of people to countries or places outside their standard environment for personal or commercial enterprise, or professional purposes. It's an activity that can be identified in every developed and developing nation. In general terms, tourism is the movement of a person from one location to another in order to experience the harmony of that location or to have fun. Traveling can be an effortful and challenging endeavor, and not everybody is ready to travel. However, at the same time, it's an enjoyable activity that takes your fatigue away. Traveling provides flavor to life as you travel to completely different places that have unique cultures and modes of operation. It's a good way to learn about the culture and tradition of a place. Additionally, for many regions, tourism is their main source of economic benefit.

With the current state of the tourism industry, the problems of tourism may also far exceed the benefits of a country. As a result, carbon monoxide

Sustainability in Marketing Practice: Strategies for Industry 4.0.
Subrata Chattopadhyay, Udit Chawla, (Eds.)

emissions may rise sharply due to increased airline and car use, negatively impacting the Indian ecology. With high foot traffic and human interaction, many countries with old remains or natural attractions are at risk of decay or degradation. Moreover, when there is an overabundance of people, plants and wildlife may decline in certain locations or alter their development and migration patterns. The results of these studies indicated that tourist society participation in tourism activities in India involved local society directly and indirectly. Additionally, ecotourism is conducted at the expense of the local society's lifestyle. Local societies have been required to learn about ecotourism, especially about the uniqueness of tourist sites, lifestyles, beliefs, and customs of local societies. It can be concluded that it's very important to pay attention to ecotourism in social-cultural aspects as it will have a negative impact on the overall sustainability of tourist destinations.

This study is undertaken to understand consumers' opinions regarding ecotourism, with respect to demographic factors. The aim is to understand how different demographic factors affect satisfaction and preference toward the ecotourism industry. The data collection point has been taken as Howrah and Kolkata. Both of these cities are filled with places of great historical and monumental importance; therefore, they make them the best fit for the point of the collection of data.

14.1 INTRODUCTION

As much as the tourism industry advantages the financial system of numerous national locations and is one of the most essential industries in the world, it is also the most responsible for the damages caused to the nature and environment, as a result of the overconsumption of natural resources like water, energy, and raw production materials. In this situation, visitors are more responsible for environmental issues because of tourism, so businesses need to choose a respectable way for the process, along with green marketing practices (Tilikidou et al., 2015). To popularize ecotourism, a green orientation of marketing in hospitality needs to be addressed to the consumers, to introduce the full green practice and employ them to eco-friendly activities (Wu et al., 2019; Chang et al., 2018, 2019).

Ecotourism is described as traveling to quite steady or clean areas with the particular goal of observing, favoring, and loving the heritage and its wild flora and animals, as well as any existing educational presentations, is one of the fastest-growing tourism sectors globally and it creates a profitable sustainable service practice environment (Hultman, Kazeminia, &

Ghasemi, 2015). It's for the convention of traveling to relatively less popular natural locations, to evaluate and experience the natural environment, gather information about wildlife, and experience local traditions in real surroundings while preserving the environments of the locations (Lee & Jan, 2019). Ecotourism has been widely supported by the institutions of protected areas in many countries that encourage the sustainable growth of tourism (Buckley et al., 2008). Sustainable tourism is an important component of the global sustainable enhancement of the country with the awareness of the impacts resulting from mass tourism, many new forms of alternative tourism have been conceptualized which include community-based tourism, fair-trade tourism, and pro-poor tourism (Lebon, 2009).

Ecotourism has received growing attention in recent years, not only as an alternative to mass tourism but as a means of economic development and environmental protection (Schaller, 2010). Techniques to enhance land for ecotourism need thorough planning and incremental enclosure of land with a focus on ecological sustainability. To reduce negative impacts, ecotourism development should be managed and adapted to the natural values and environmental knowledge of a particular area (Li et al., 2012). As a result, many writers argue that the development of ecotourism can only be achieved through the involvement of local authorities and the people in the management process (Ramosa & Prideauxa, 2014; Wishitemi et al., 2015). Researchers have measured ecotourism as a means of reducing environmental and socioeconomic issues and as a sustainable development tool in economically responsive areas. However, many efforts to study the literature on ecotourism are rarely focused and fragmented, resulting in a lack of comprehensive understanding of this topic, which is most critical in the present situation because concepts such as ecotourism are multidisciplinary (Boley, Maruyama, & Woosnam, 2015).

While an individual discusses ecotourism, they are speaking about activities which have a large impact on the society or area to which they're visiting. Diminishing the carbon footprint or reducing the waste they create while traveling by selecting reusable stuff like a refillable water bottle, all play a part in decreasing their overall impact. Despite the fact that the truly sustainable vacation is about assisting individuals and infrastructure long after they've departed: following an area socially, economically, and environmentally (Anon, 2010). By seeking out ecotourism travel knowledge, people will also discover a unique opportunity to connect more deeply with the people and places they visit, understanding the culture, people, wildlife, and various attractions each destination has to offer more responsibly.

Travelers are interested in natural landscapes which include large biodiversity. Travelers in the coastal area get pleasure from swimming in clean waters along with the fish and coral reefs and observing whales and seabirds. Others go on safaris to study wildlife. These kinds of activities involve an untouched and healthy environment. National parks rely on a functioning environment to provide travelers with recreation, education, culture, and fun (Melita & Mendlinger, 2013). Ecotourism is specifically designed to protect natural areas and therefore plays a part in the conservation of biodiversity. It concentrates especially on understanding and learning about the environment, its landscape, plants, fauna, and their habitats, as well as cultural artifacts from the region. In ecotourism, learning the major concern is the environment and its conservation (CBD Secretariat, 2013).

The money that travelers use can provide nature, society, and culture in the form of protected areas and other attractions. Sustainable tourism can also create a society proud of upholding and sharing their traditions, knowledge, and art, which contribute to the sustainable use of limited biodiversity. Tourism can have a variety of negative impacts on biodiversity, mainly when there isn't enough management. Reckless and unsustainable tourism can harm nature through environmental damage, overexploitation of local resources, waste and pollution, infrastructure development, and greenhouse gas emissions. Travelers expect a clean environment and they will not return to spoiled or ruined destinations, which will result in economic losses (Nasi & Belair, 2010).

It helps in social development by giving an alternative source of livelihood to the local community that is more sustainable. Many people view ecotourism as a practical way to protect the natural environment and provide social and economic benefits to the local areas. Ecotourism includes a range of nature-based activities that promote tourist appreciation and understanding of natural and cultural heritage and are aimed to be ecologically, economically, and socially sustainable. As a result, ecotourism is recognized as an alternative form of sustainable development. Its purpose is to conserve resources, particularly biological diversity, and maintain the sustainable use of resources, which can bring a natural experience to tourists, protect the natural environment, and provide benefits (Drive, 2001).

14.2 LITERATURE REVIEW

Ecotourism is considered as a subsection of sustainable tourism and is supposed to have the potential to become an important part of the tourism industry (Lai & Shafer, 2005). Research has suggested that environmentally friendly travelers are more interested in seeking information about the destination and whether it is ecologically friendly or not (Dolnicar & Long, 2009). Travelers are eager to contribute to environmental management and cultural preservation at the destination (Buhalis & Law, 2008). Ecological management and cultural preservation are the main components of ecotourism (Das & Hussain, 2016), highlighting the fact that travelers use information to search for facts related to ecotourism.

Ecotourism involves environment-based activities and has a significant impact on the environment (Singh et al., 2015). It also has a range of benefits for the local community (Doroban et al., 2012). According to the World Tourism Organization, ecotourism is characterized by being small-scale and involving local industries, having a strong educational focus, reducing negative environmental impacts, and promoting socio-economic progress in natural areas (UNWTO, 2020; Fennell, 2020). In the green economy, ecotourism marketing plays a crucial role in providing consumers with important information about the level of responsibility taken by visited destinations and aims to meet the demands of eco-conscious travelers (Gheorghe et al., 2016).

Eco-tourism has a unique relationship with the environment as the natural and cultural environment is closely linked to tourism. Destinations with beautiful landscapes, diverse biodiversity, unique heritage sites, and vibrant indigenous cultures are competitive in the tourism market. However, the tourism sector consumes resources, generates waste, and contributes to large-scale infrastructure development, leading to environmental degradation (Sharpley, 2003). Water resources are particularly affected, with hotels and recreational facilities overusing water for the convenience of tourists, resulting in wastage and degradation (Sunlu et al., 2003).

Eco-tourism also plays a role in preserving historical and cultural traditions. The socio-cultural aspects of tourism are important as they can be assets or liabilities. Countries strive to protect their cultural heritage, including historical monuments, arts, and crafts, to attract tourists. It also promotes understanding of the host country's customs, festivals, music, cuisine, religion, language, etc., which helps to preserve cultural identity (Rajagopalan et al., 2009). Communication between tourists and local residents involves a learning process in which they appreciate the culture, ideas, and values of different countries (Nyaupane, 2006).

Environmental challenge, also referred to as a pro-environment outlook or environmental fact, refers back to the records of a man or woman about the significance of protective the environment. Motivational paradigms recommend a beneficial structure for the attempting to find behaviors, particularly environmentally responsible behaviors (Chiu et al., 2014a). Research takes into account environmentally responsible behaviors as the conclusion of the environmental view (Kang & Moscardo, 2006).

Destination picture plays a crucial position in travel desire. It persuades the selection-making method with the aid of connecting to a spread of destinations and additionally instances after-choice-making behaviors regarding participation, assessment, and future behavioral purpose (Chen & Tsai, 2007). Whilst a tourist has a fine reflection of a place, the hazard of journeying that place increases. Usually, if an eco-vacation spot seems to be attractive to a vacationer, the opportunity of traveling to the eco-destination turns out to be better.

Most significantly, tourism has fashioned a multiplier motive on India's socio-economic development thru infrastructure growth, job advent, and talent improvement, amongst others (Rajasekharan, 2006). Financial effects including employment opportunities, prosperity with inside the common of alive, and company improvement impact social rewards which encompass allowing of persuaded governments and a flourishing in ecological mindfulness and lastly results in environmental conservation (Sarkar & Sinha, 2015). Tourism is a service-creating Indian enterprise that offers employment possibilities to a massive cross-segment of inexpert, semi-expert, and highly professional professionals. Researchers determined that urban societies were cost-effectively better off while related to their countryside counterparts. The massive difference among the 2 societies turned into special by modifications in wealth, human capital, and employment degrees, and so condensed financial increase inside the rural societies (Chalip & Costa, 2012). Authors have referred to the monetary effect of tourism as the maximum essential allowing issue for the host Center of the population to grow to be self-sustainable (Nunkoo & Gursoy, 2012).

Hypotheses Formulation Based on the research insights in the literature review, the next research theories can be developed:

- **H1:** Ecotourism is dependent on the gender of tourists.
- **H2:** Ecotourism is dependent on the age group of tourists.
- **H3:** Ecotourism is dependent on the monthly income of tourists.
- **H4:** Ecotourism is dependent on the educational qualification of tourists.
- **H5:** Ecotourism is dependent on the occupation of tourists.

14.3 OBJECTIVE OF THE STUDY

The objective of the study is to study the relationship between demographic factors and satisfaction among customers towards ecotourism.

14.4 RESEARCH METHODOLOGY

This research is complete with the help of structured questionnaires, 200 samples were calm from the metro city, Kolkata through one-to-one interaction with the tourists related to Eco-Tourism. The main facts for the education were composed of men, and women, of different age groups, belonging to different income supports, from various occupation backgrounds like business, service, and others. Each parameter study for the survey was intended using five points like a scale (from 1-strongly affect, to 5-strongly agree). We used the chi-square test to find out the significant relationship of the demographic factors to the satisfaction among tourists towards Eco-Tourism.

14.5 DATA ANALYSIS

We have used the Chi-square test using SPSS 21.0 to prove the hypothesis given below:

- **H0:** Eco-tourism is independent of the gender of the tourists. Eco-Tourism is dependent on the gender of the tourists.
- **H0:** Eco-Tourism is independent of the age group of the tourists. Eco-Tourism is dependent on the age group of the tourists.
- **H0:** Eco-Tourism is independent of the monthly income of the tourists. Eco-Tourism is dependent on the monthly income of the tourists.
- **H0:** Eco-Tourism is independent of the educational qualification of the tourists. Eco-Tourism is dependent educational qualification of the tourists.
- **H0:** Eco-Tourism is independent of the occupation of the tourists. Eco-Tourism is dependent on the occupation of the tourists.

14.5.1 SATISFACTORY VARIABLES

- **V6:** Ecotourism help me explore cultural places and know there historical stories.

- **V10:** Ecotourism builds employment opportunity for the local community.
- **V19:** Ecotourism helps spread awareness about protecting the environment.

14.6 RESULT AND ANALYSIS

14.6.1 GENDER

From Table 14.1, we see that the chi-square value for V6 is 4.193[a] and the symposia are 0.381 which is more than 0.05. Similarly, the chi-square value for V10 is 3.167[a] and the asymp sig is 0.530 which is also more than 0.05, and the chi-square value for V19 is 6.772[a] and asymp sig is 0.148 which is also more than 0.05. This implies that the null hypothesis is accepted, and the alternate hypothesis is rejected. Thus, gender has a significant association with tourist satisfaction with respect to variables V6, V10, and V19. Similarly, it can be seen that the Contingency Coefficient of variable V6 is 0.143 and for variable V10 is 0.125, and for variable V19 is 0.181 this indicates the relationship between gender and tourist satisfaction is low with the variable.

TABLE 14.1 Chi-Square Tests and Symmetric Measures of Gender

		Value	df	Asymp. Sig. (2-Sided)			Value	Approx. Sig.
V6	Pearson chi-square	4.193a	4	0.381	Nominal by nominal	Contingency coefficient	0.143	0.381
	Likelihood ratio	5.561	4	0.234				
	No. of valid cases	200	–	–	No. of valid cases		200	–
V10	Pearson chi-square	3.167a	4	0.530	Nominal by nominal	Contingency coefficient	0.125	0.530
	Likelihood ratio	4.552	4	0.336	–	–	–	–
	No. of valid cases	200	–	–	N of valid cases		200	–
V19	Pearson chi-square	6.772a	4	0.148	Nominal by nominal	Contingency coefficient	0.181	0.148
	Likelihood ratio	8.075	4	0.089				
	No. of valid cases	200	–	–	No. of valid cases		200	–

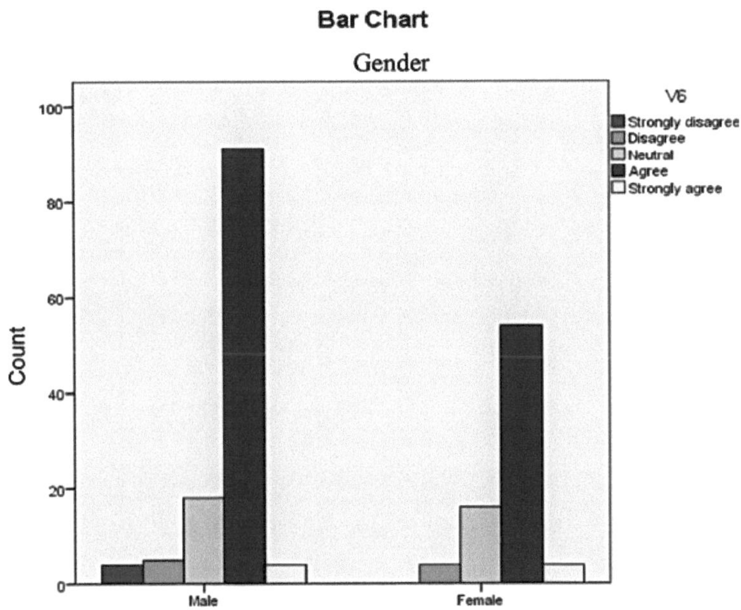

FIGURE 14.1 Chi square graphs (Image 1).

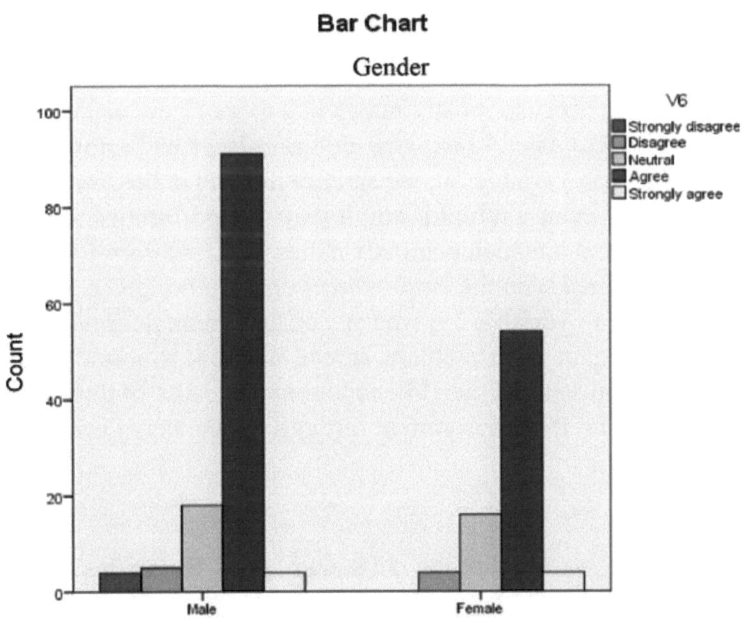

FIGURE 14.2 Chi square graphs (Image 2).

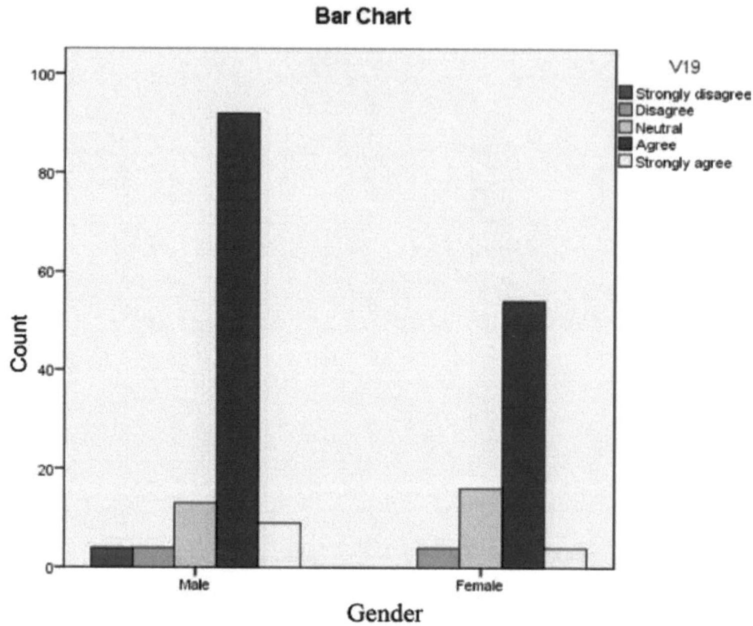

FIGURE 14.3 Chi square graphs (Image 3).

Concerning variable 6, we can see that the males are more satisfied with ecotourism because helps them explore cultural places and know their historical stories, because mostly males like to tour new places and because of that they are interested to explore cultural places and know their historic places. Similarly in variable 10, we can see that the males are more satisfied with ecotourism because it builds employment opportunities for the local co. After all, because in our society mostly males work and earn for family's mile and they are satisfied with the employment opportunity given by ecotourism. Correspondingly to variable 19, we can see that the males are more satisfied with ecotourism that it helps them spread awareness about protecting the environment because most men like to tour and because of that they are more interested in saving the environment through the means of ecotourism.

14.6.2 AGE

From Table 14.2, we see that the chi-square value for V6 is 60.681[a] and the symposia is 0.000 which is less than 0.05. Similarly, the chi-square value for V10 is 118.572a and the asymp sig is 0.000 which is also less than 0.05, and, that the chi-square value for V19 is 46.742a and the asymp sig is 0.000 which is also less than 0.05. This implies that the null hypothesis is rejected, and the alternate hypothesis accepted. So, we can say that age has a significant association with tourist satisfaction with respect to variables V6, V10, and V19. Similarly, it can be seen that the Contingency Coefficient of variable V6 is 0.482 and for variable V10 is 0.610, and for variable V19 is 0.435 this indicates the relationship between age and tourist satisfaction is moderate with the variable.

TABLE 14.2 Chi-Square Tests and Symmetric Measures of Age

Variables	Tests	Value	df	Asymp. Sig. (2-Sided)			Value	Approx. Sig.
V6	Pearson chi-square	60.681a	16	0.000	Nominal by nominal	Contingency coefficient	0.482	0.000
	Likelihood ratio	80.193	16	0.000				
	No. of valid cases	200	–	–	No. of valid cases		200	0
V10	Pearson chi-square	118.572a	16	0.000	Nominal by nominal	Contingency coefficient	0.610	0.000
	Likelihood ratio	130.313	16	0.000				
	No. of valid cases	200	–	–	No. of valid cases		200	–
V19	Pearson chi-square	46.742a	16	0.000	Nominal by nominal	Contingency coefficient	0.435	0.000
	Likelihood ratio	59.404	16	0.000				
	No. of valid cases	200	–	–	No. of valid cases		200	–

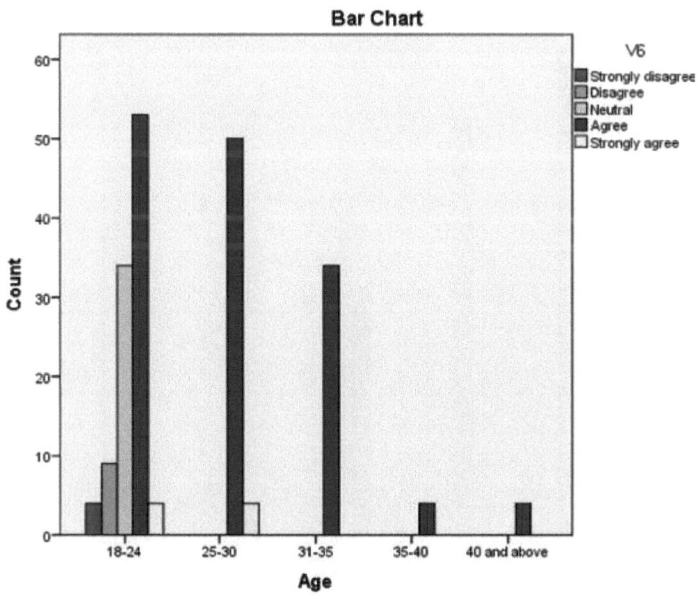

FIGURE 14.4

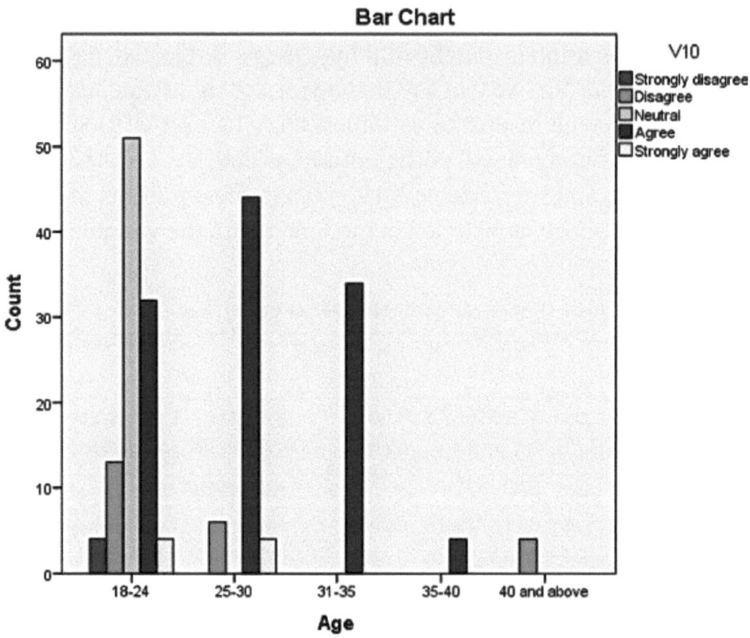

FIGURE 14.5

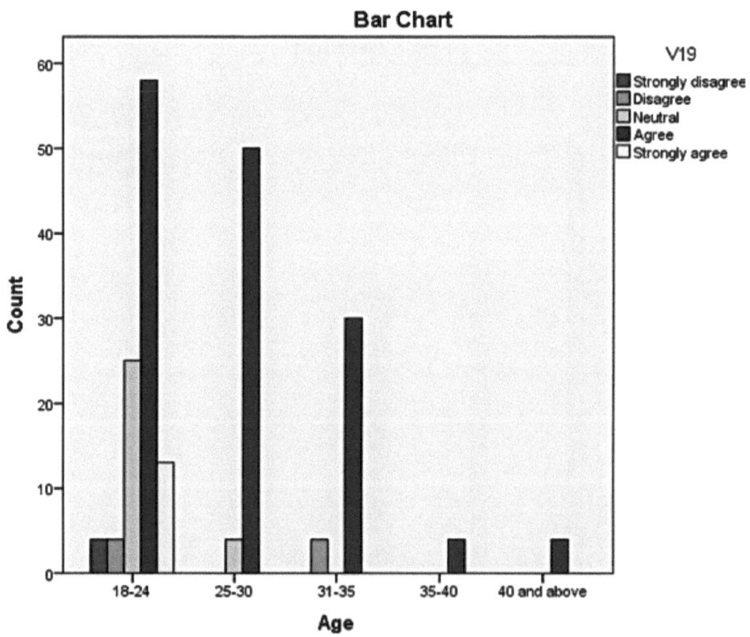

FIGURE 14.6

Concerning variable 6, we see that younger tourist 18–24 is Evid satisfied with the ecotourism that helps them explore cultural places and their historical stories because nowadays youth are interested to explore the country's cultures and getting knowledge of its history. Concerning variable 10, we can see that the adult tourists of age 25–30 are identified with the ecotourism that it builds employment opportunities Cal community because the adults of age 25–30 are job working people and many adults are unemployed so because of ecotourism the adults of age 25–30 are getting job opportunities. Concerning can see that younger tourist 18–24 is more satisfied with ecotourism that it helps to spread awareness about protecting the environment because nowadays youth are aware of the consequences of global warming and that is why they are spreading awareness to protect the environment through the means of ecotourism.

14.6.3 INCOME

From Table 14.3, we see that the Chi-square value for V6 127.363[a] and the asymp sig is 0.00 which is less than 0.05. Similarly, the chi-square value for V10 is 70.376[a] and the asymp sig is 0.000 which is also less than 0.05, and, that the chi-square value for V19 is 84.437[a] and the asymp sig is 0.000 which is also less than 0.05. This implies that the null hypothesis is rejected, and the alternate hypothesis accepted. So, we can say that income has a significant association with tourist satisfaction with respect to variables V6, V10, and V19. Similarly, it can be seen that the Contingency Coefficient of variable V6 is 0.624, and for variable V6 is 0.510, and for variable V19 is 0.545 which indicates the relationship between income and tourist satisfaction is moderate with the variable.

TABLE 14.3 Chi-Square Tests and Symmetric Measures of Income

		Value	df	Asymp. Sig. (2-Sided)			Value	Approx. Sig.
V6	Pearson chi-square	127.363a	16	0.000	Nominal by nominal	Contingency coefficient	0.624	0.000
	Likelihood ratio	96.023	16	0.000				
	N of valid cases	200	–	–	No. of valid cases		200	–
V10	Pearson chi-square	70.376a	16	0.000	Nominal by nominal	Contingency coefficient	0.510	0.000
	Likelihood ratio	91.986	16	0.000				
	No. of valid cases	200	–	–	No. of valid cases		200	–
V19	Pearson chi-square	84.437a	16	0.000	Nominal by nominal	Contingency coefficient	0.545	0.000
	Likelihood ratio	81.154	16	0.000				
	No. of valid cases	200	–	–	No. of valid cases		200	–

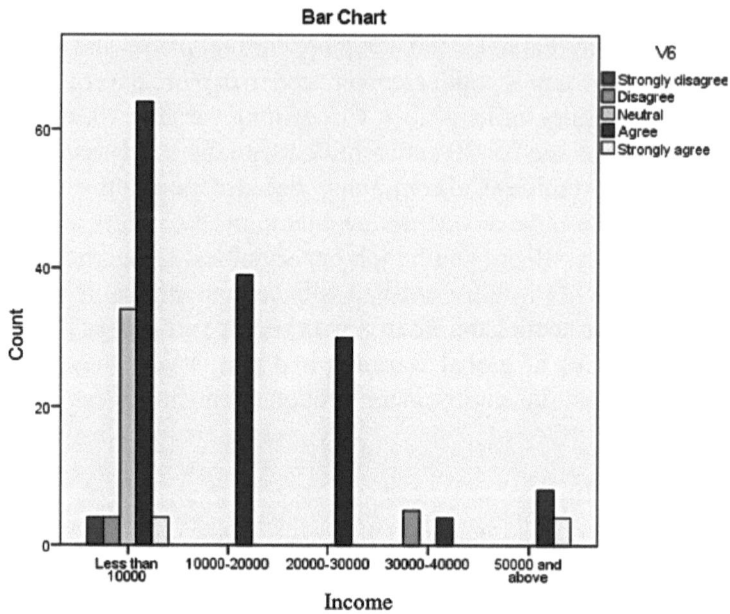

FIGURE 14.7

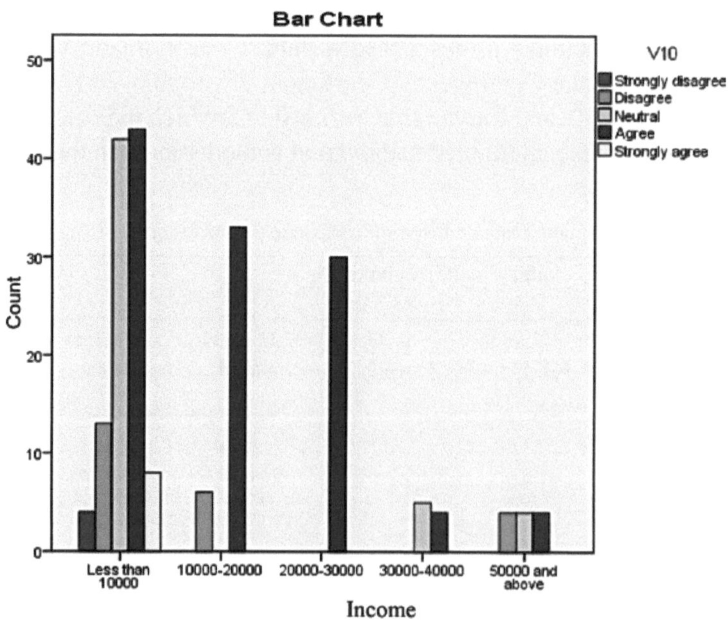

FIGURE 14.8

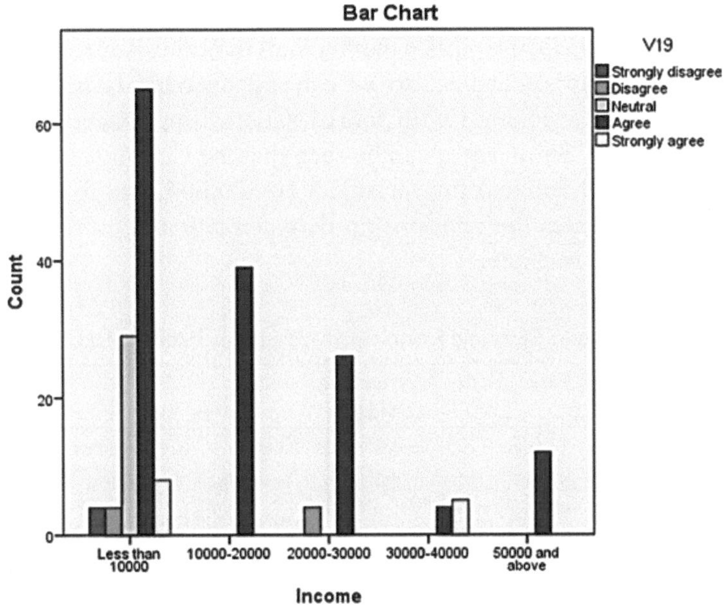

FIGURE 14.9

Concerning variable 6, we can see that the individuals whose income is less than 10,000 are more satisfied with ecotourism helping explore cultural places and know their historical stories because they want to know about their regional culture and its history so that in future, they can get a job of a tourist guide that will help him/her and its family. Concerning variable 10, we can see that the individuals whose income is less than 10,000 are highly satisfied with ecotourism building employment opportunities, because income like 10,000 does not fulfill one's family's daily needs and because of that, they are searching for new employment opportunities. Concerning variable 19, we can see that the individuals whose income is less than 10,000 are more satisfied with ecotourism to helping spread awareness to protect the environment because they are not rich so most of them live in the countryside and they know how pollution affects them so that they can spread awareness through ecotourism to tourists to not to pollute land, water, and air and use biodegradable products.

14.6.4 EDUCATIONAL QUALIFICATION

From Table 14.4, we see that the chi-square value for V6 95.106a is and the asymp sig is 0.000 which is less than 0.05. Similarly, the chi-square value for V10 is 114.637[a] and the asymp sig is 0.000 which is also less than 0.05, and, that the chi-square value for V19 is 99.426[a] and the asymp sig is 0.000 which is also less than 0.05. This implies that the null hypothesis is rejected, and the alternate hypothesis is accepted. So, we can say that educational qualification has a significant association with tourist satisfaction concerning variables V6, V10, and V19. Similarly, it can be seen that the Contingency Coefficient of variable V6 is 0.568 and for variable V10 is 0.604, and for variable V19 is 0.576 this indicates the relationship between age and tourist satisfaction moderate with the variable.

TABLE 14.4 Chi-Square Tests and Symmetric Measures of Educational Qualification

		Value	df	Asymp. Sig. (2-Sided)			Value	Approx. Sig.
V6	Pearson chi-square	95.106a	12	0.000	Nominal	Contingency	0.568	0.000
	Likelihood ratio	51.817	12	0.000	By nominal	Coefficient	–	–
	No. of valid cases	200	–	–	No. of valid cases		200	–
	Pearson chi-square	114.637a	12	0.000	Nominal by nominal	Contingency coefficient	0.604	0.000
V10	Likelihood ratio	72.514	12	0.000				
	No. of valid cases	200	–	–	No. of valid cases		200	–
	Pearson chi-square	99.426a	12	0.000	Nominal by nominal	Contingency coefficient	0.576	0.000
V19	Likelihood ratio	55.555	12	0.000				
	No. of valid cases	200	–	–	No. of valid cases		200	–

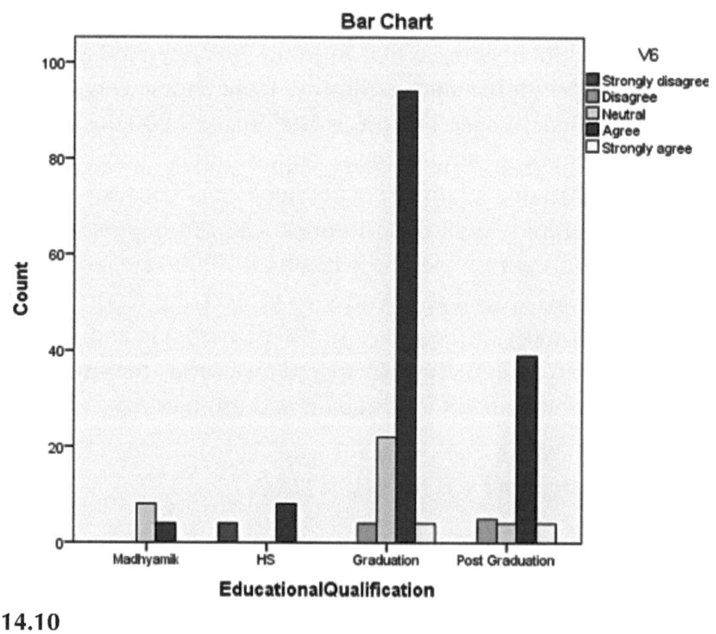

FIGURE 14.10

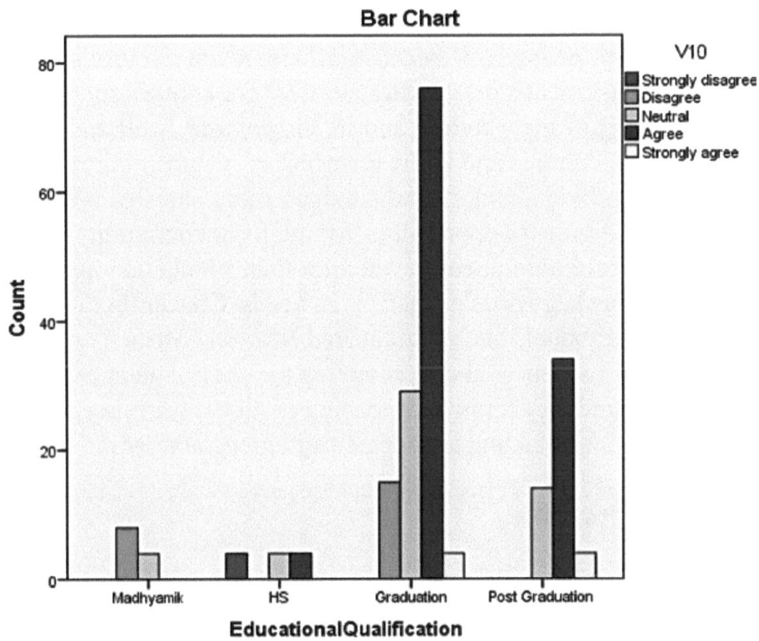

FIGURE 14.11

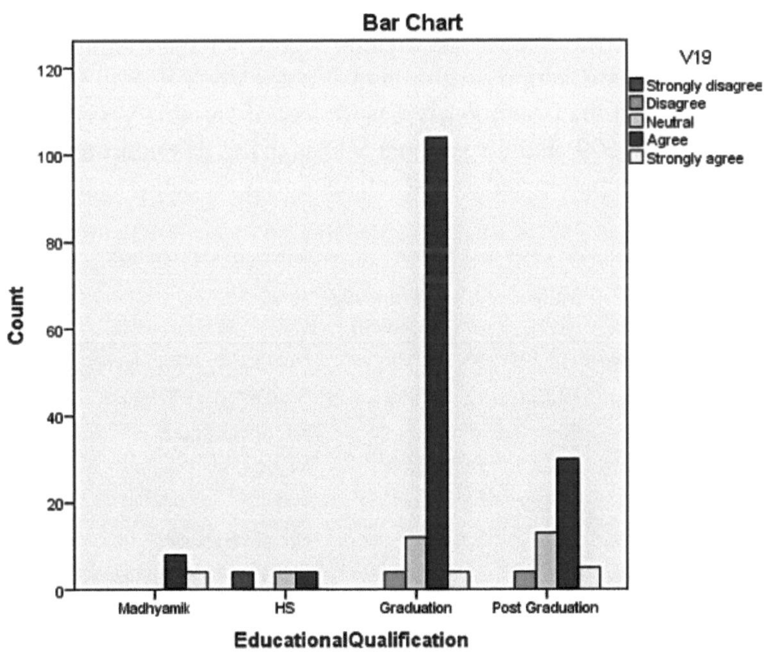

FIGURE 14.12

Concerning variable 6, we can see that the individuals that graduated are more satisfied with ecotourism because it helps them explore cultural places and know their historical stories. After all, they are enough educated and they know the meaning of their culture and its history very well and they want to explore more new cultures and know their history. Concerning variable 10, we can see that the individuals that graduated are more satisfied with ecotourism that it builds employment opportunities for the local community because most graduated students remain unemployed after their graduation and even if they get a job then its package does not satisfy its needs. Concerning variable 19, we can see that the individuals that are graduated are more satisfied with ecotourism by helping spread awareness about protecting the environment because they are enough educated and they know the meaning of global warming and they know how to prevent it by spreading awareness through ecotourism.

14.6.5 OCCUPATION

From Table 14.5, we see that the chi-square value for V6 is 112.295[a] and the asymp sig is 0.000 which is less than 0.05. Similarly, the chi-square value for V10 is 118.206[a] and the asymp sig is 0.000 which is also less than 0.05, and, that the chi-square value for V19 is 43.277[a] and the asymp sig is 0.000 which is also less than 0.05. This implies that the null hypothesis is rejected, and the alternate hypothesis is accepted. So, we can say that occupation has a significant association with tourist satisfaction concerning variables V6, V10, and V19. Similarly, it can be seen that the Contingency Coefficient of variable V6 is 0.600 and for variable 106 is 0.609, and for variable V17 is 0.422 this indicates the relationship between occupation and tourist satisfaction is moderate with the variable.

TABLE 14.5 Chi-Square Tests and Symmetric Measures of Occupation

		Value	df	Asymp. Sig. (2-Sided)			Value	Approx. Sig.
V6	Pearson chi-square	112.295[a]	12	0.000	Nominal by nominal	Contingency coefficient	0.600	0.000
	Likelihood ratio	80.193	12	0.000				
	No. of valid cases	200	–	–	No. of valid cases		200	–
V10	Pearson chi-square	118.206[a]	12	0.000	Nominal by nominal	Contingency coefficient	0.609	0.000
	Likelihood ratio	58.620	12	0.000				
	No. of valid cases	200	–	–	No. of valid cases		200	–
V19	Pearson chi-square	43.277a	12	0.000	Nominal by nominal	Contingency coefficient	0.422	0.000
	Likelihood ratio	39.701	12	0.000				
	No. of valid cases	200	–	–	No. of valid cases		200	–

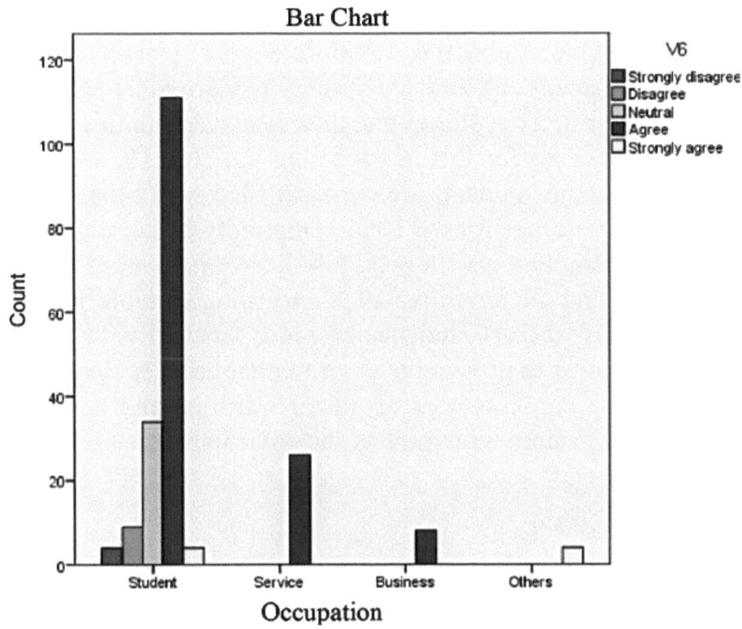

FIGURE 14.13

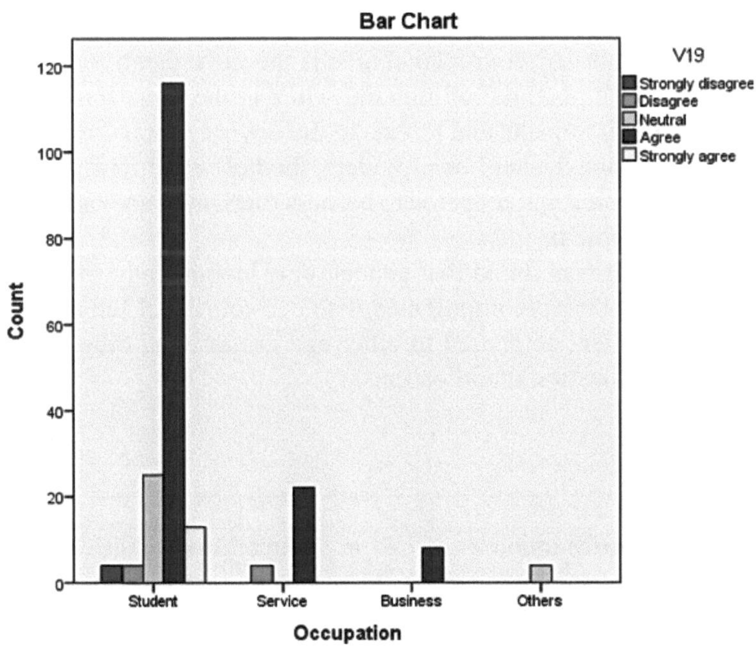

FIGURE 14.14

Concerning variable 6, we can see that the students are more satisfied with ecotourism which helps them explore cultural places and know their historical stories because as they are young, they are more enthusiastic and energetic and interested in cultures and their historical studies whereas other people don't have time for this. Concerning variable 10, we can see that the individuals who are students are more satisfied with ecotourism to build employment opportunities for the local community because they know that after their education, they want a good job. Ecotourism helps spread awareness about protecting the environment. Concerning variable 19, we can see that the individuals who are students are more satisfied with ecotourism to help spread awareness to protect the environment because they are educated and they know the consequences of global warming and how to tackle it through spreading awareness to protect the environment.

14.7 CONCLUSION

The program towards ecotourism is right away a threat and an opportunity to create more sustainable tourism: by way of diverting traveler stream to vow the carrying capacity of any destination isn't exceeded: through planning for renewal of natural resources; and by generating awareness within the local community wherein they may be set and forearmed to address the harmful impact of mass tourism. A crucial course is the most creative way to maximize the economic potential of tourism, while at the same time minimizing the negative social impacts and threats to the environment. Only ecotourism wherein the tourists, the service providers, the host community and authorities are well informed and prepared to harness tourism as an engine of growth can yield sustainable results.

In our study, it was found that gender, age, income, education qualification, and occupation have importance with eco-tourism. Men are seen to be happier than women, compared to other age groups, and other occupations or with higher education qualifications.

KEYWORDS

- **customer satisfaction**
- **demographic factors**
- **ecotourism**
- **educational qualification**
- **satisfactory variables**
- **sustainability**

REFERENCES

Anon., (2010). In: Ben, S., (ed.), *The Importance of Education in Ecotourism Ventures.* Substantial Research Paper American University.

Boley, B. B., Maruyama, N., & Woosnam, K. M., (2015). Measuring empowerment in an eastern context: Findings from Japan. *Tourism Management, 50*, 112–122.

Buckley, R., Cater, C., Linsheng, Z., & Chen, T., (2008). Shengtailuyou: Cross-cultural comparison in ecotourism. *Annals of Tourism Research, 35*(4), 945–968.

Buhalis, D., & Law, R., (2008). Progress in information technology and tourism management: 20 years on and 10 years after the internet – the state of e-Tourism research. *Tourism Management, 29*(4), 609–623. https://doi.org/10.1016/j.tourman.2008.01.005.

CBD Secretariat, (2013). *Biodiversity for Sustainable Development, 105*(12), 2013.

Chalip, L., & Costa, C. A., (2012). Clashing worldviews: Sources of disappointment in rural hospitality and tourism development. *Hospitality and Society, 2*(1), 25–47.

Chen, C. F., & Tsai, D., (2007). How destination image and evaluative factors affect behavioral intentions? *Tourism Management, 28*(4), 1115–1122.

Chiu, Y. T. H., Lee, W. I., & Chen, T. H., (2014a). Environmentally responsible behavior in ecotourism: Antecedents and implications. *Tourism Management, 40*, 321–329.

Das, D., & Hussain, I., (2016). Does ecotourism affect economic welfare? Evidence from Kaziranga National Park, India. *Journal of Ecotourism, 15*(3), 241–260. https://doi.org/10.1080/14724049.2016.1192180.

Dolnicar, S., & Long, P., (2009). Beyond ecotourism: The environmentally responsible tourist in the general travel experience. *Tourism Analysis, 14*(4), 503–513. https://doi.org/10.3727/108354209X12596287114291.

Doroban¸tu, M. R., & Nistoreanu, P., (2012). Hospitality–a component of Romanian rural tourism products. *Cactus Tour. J., 3*, 17–23.

Drive, C., (2001). *Case Study: Tourism and Biodiversity (Ecotourism – A Sustainable Development Tool, a Case for Belize).*

Fennell, D. A., (2020). Ecotourism. *International Encyclopedia of Human Geography* (pp. 51–55). Department of Tourism and Environment, Brock University: St. Catharines, ON, Canada.

Gheorghe, A., & Pârvu, F., (2016). Aspects regarding green economy's. *Bus. Excell. Manag., 6*, 21–35.

https://www.unwto.org/sustainabledevelopment/ecotourism-and-protected-areas (accessed on 15 February2020).

Hultman, M., Kazeminia, A., & Ghasemi, V., (2015). Intention to visit and willingness to pay a premium for ecotourism: The impact of attitude, materialism, and motivation. *Journal of Business Research, 68*(9), 1854–1861.

Kang, M., & Moscardo, G., (2006). Exploring cross-cultural differences in attitudes towards responsible tourist behavior: A comparison of Korean, British and Australian tourists. *Asia Pacific Journal of Tourism Research, 11*(4), 303–320.

Lai, P. H., & Shafer, S., (2005). Marketing ecotourism through the internet: An evaluation of selected ecolodges in Latin America and the Caribbean. *Journal of Ecotourism, 4*(3), 143–160. https://doi.org/10.1080/JET.v4.i3.pg143.

Lebon, C., (2009). *Tourism and Sustainability: Development, Globalization and New Tourism in the Third World* (3rd edn.) Routledge: Oxon, UK. [CrossRef].

Lee, T. H., & Jan, F. H., (2019). Can community-based tourism contribute to sustainable development? Evidence from residents' perceptions of sustainability. *Tourism Management, 70*, 368–380.

Li, B., Zhang, F., Zhang, L., Huang, J., Jin, Z., & Gupta, D. K., (2012). Comprehensive suitability evaluation of tea crops using GIS and a modified land ecological suitability evaluation model. *Pedosphere, 22*, 122–130.

Melita, A. W., & Mendlinger, S., (2013). *The Impact of Tourism Revenue on the Local Communities' Livelihood: A Case Study of Ngorongoro Conservation Area, Tanzania* (pp. 117–126).

Nasi, R., & Belair, C., (2010). *Tropical Forest Management and Sustainability Issues.* Available: http://www.ffpri.affrc.go.jp/event/2010/20100426cbdcop10/documents/12nasi.pdf; http://www.academia.edu/attachments/8155711/download_file; http://www.academia.edu/132159/Tropical_forest_management_and_sustainability_issues; http://s3.amazonaws.com/ (accessed on 6 December 2023).

Nunkoo, R., & Gursoy, D., (2012). Residents' support for tourism: An identity perspective. *Annals of Tourism Research, 39*(1), 243–268.

Nyaupane, G. P., Morais, D. B., & Dowler, L., (2006). The role of community involvement and number/type of visitors on tourism impacts A controlled comparison of Annapurna, Nepal and Northwest Yunnan, Nepal. *Tourism Management, 27*, 1373–1385.

Rajagopalan, P., & Sebastian, L. M., (2009). Socio-cultural transformations through tourism: A comparison of residents' perspectives at two destinations in Kerala, India. *Journal of Tourism and Cultural Change, 7*(1), 5–21. https://doi.org/10.1080/14766820902812037.

Rajasekharan, P. K., (2006). *Structure of the Labor Market in the Tourism Sector of Kerala.* Unpublished Thesis, Department of Economics, University of Kerala.

Ramosa, A. M., & Prideauxa, B., (2014). Indigenous ecotourism in the Mayan rainforest of Palenque: Empowerment issues in sustainable development. *Journal of Sustainable Tourism, 22*(3), 461–479.

Sarkar, R., & Sinha, A., (2015). The village as a social entrepreneur: Balancing conservation and livelihoods. *Tourism Management Perspectives, 16*, 100–106. doi: 10.1016/j.tmp.2015.07.006.

Schaller, D., (2010). *Indigenous Ecotourism and Sustainable Development: The Case of Río Blanco, Ecuador.* http://www.eduweb.com/schaller/RioBlancoSummary.html (accessed on 6 December 2023).

Sharpley, R., (2003). *Tourism, Tourists, and Society*. Elm Publications, Huntingdon.

Singh, S., Dash, T. R., & Vashko, I., (2015). Tourism, ecotourism, and sports tourism: The framework for certification. *Mark. Intell. Plan., 34*, 236–255.

Sunlu, U., (2003). *Environmental Impacts of Tourism* (pp.263–270). http://om.ciheam.org (accessed on 6 December 2023).

Tilikidou, I., & Delistavrou, A., (2015). Preferring green and rejecting "unethical" hotels. *Euro. Med. J. Bus., 10*, 285–296.

UNWTO, (2020). Ecotourism and Protected Areas. Available online:

Wishitemi, B. E. L., Momanyi, S. O., Ombati, B. G., & Okello, M. M., (2015). The link between poverty, environment, and ecotourism development in areas adjacent to Maasai mara and Amboseli protected areas, Kenya. *Tourism Management Perspectives, 16*, 306–317.

Wu, H. C., Cheng, C. C., & Ai, C. H., (2019). What drives green experiential loyalty towards green restaurants? *Tour. Rev.*, 1–20.

Wu, H. C., Wei, C. F., Tseng, L. Y., & Cheng, C. C., (2018). What drives green brand switching behavior? *Mark. Intell. Plan., 36*, 694–708.

CHAPTER 15

Why Do Consumers Behave… The Way They Do! (A Bird's-Eye View on the Heuristics and Biases in Consumerism)

RAJIB DUTTA

Amity Business School, Amity University, Kolkata, West Bengal, India

ABSTRACT

In behaviorism, any form of mental ability that prompts the consumer to decide and think rapidly without any overriding input can be termed as heuristics. When consumers form judgments and make decisions without spending too much time on researching, they use mental shortcuts. Such mental shortcuts or strategies are heuristics. Knowingly or unknowingly, every consumer is seen to apply 'heuristics' in their day-to-day life – be it problem-solving or even decision-making. Hence, heuristics enable faster decision making among consumers. Since heuristics seem to work under all circumstances, they are much in demand – especially from the prospective buyers' point of view. However, at times, heuristics can also lead to systematic errors. Such systematic errors lead to cognitive biases. Consumer's decision making is intertwined with cognitive biases, thereby riddled with possible errors. Such decision-making techniques also tend to make the prospects behave irrationally. Based on available information, consumers make decisions and come up with opinionated judgments. Such judgments bear the peril of being influenced by a quintet of associated biases. The fact of the matter is that, while illusion of attention, social proof or reference, peak and effect, recency and availability are some of the commonly referred to 'heuristics,' anchoring, omission, negativity, simplicity, activity, and familiarity are some of the common biases in vogue today. The irony is,

Sustainability in Marketing Practice: Strategies for Industry 4.0.
Subrata Chattopadhyay, Udit Chawla, (Eds.)

though encountering many of these behavioral inefficiencies on a day-to-day basis, consumers still get sucked in.

15.1 INTRODUCTION

The decision making of any consumer hinges on personal or situational factors. Heuristics are rules-of-thumb to aid the consumer to arrive at a particular decision – either situational (Payne et al., 1993) or personal. Derived from the Greek language, Heuristics first found mention in cognitive psychology as – the useful shortcut, an approximation, or a rule of thumb for searching through a space of viable solutions (Reimer & Hoffrage, 2004). Availability, representativeness, and adjustment are three unique types of heuristics (Kahneman & Tversky, 1974).

Yet another important research was a study on the impact of scarcity messages on consumer behavior. Limited Quantity and Limited Time advertisements influence the consumers' intention to purchase but limited quantity messages are more effective (Aggarwal & Jong, 2011). No wonder there is a frenzied and impulse buying amongst the public for 'one-day-only' Black Friday sales (Figure 15.1).

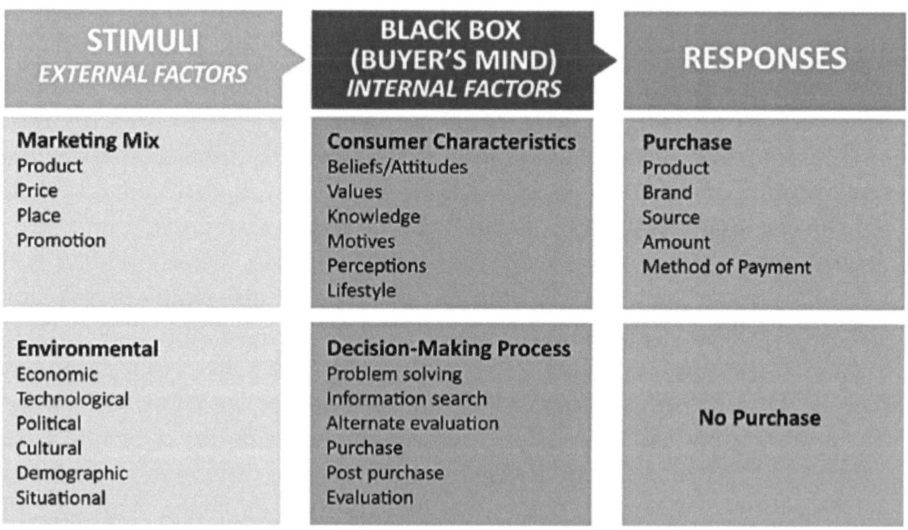

FIGURE 15.1 Black box of consumer behavior.

Source: Vinod Sonkar, https://s3-us-west-2.amazonaws.com/courses-images-archive-read-only/wp-content/uploads/sites/1505/2016/03/14215621/Black_Box_Model-1.png.

15.2 THE HEURISTICS

In behaviorism, any form of mental ability that prompts the consumer to decide and think rapidly without the baggage of any overriding knowledge can be termed as heuristics.

When, without spending sufficient time to research and analyze the facts and figures, people start firming up opinions to make decisions, they generally lean towards applying mental shortcuts or strategies, popularly known as Heuristics.

Such applications lead to lesser info-processing and decision-making time and enable smoother transitions in the flow of information and action.

Knowingly or unknowingly, every consumer is seen to take recourse to Heuristics for almost every sales/service decision made. Such strategies are also found to be in vogue and adopted by prospective buyers, as well.

15.2.1 REPRESENTATIVE HEURISTIC

Any event's probability of occurrence is subjectively determined by the degree of its aligned characteristics to the parent, and its salient features and attributes (Kahneman & Tversky, 1974). Stereotyping comes naturally to the fore when the human brain deals with probabilities.

15.2.1.1 REAL LIFE EXAMPLES

- An example of representative heuristic is to look up people who partake in purchase of lottery tickets. Despite the common knowledge that all numbers and patterns have the same statistical likelihood, owners still prefer random sequences to patterns while buying the tickets (Rachubik & Krawczyk, 2019).
- Contextual information often leads nurses in hospitals to ignore, overlook or misappropriate prominent physiological symptoms in patients (Brannon & Carson, 2003).
- During construction project home inspections, the inspector is often seen to give judgments on the entire edifice basis a sample. All this to speed up the process (Sprinkle, 2019).
- Regardless of the (in)existence of any trading regulation, the 'heuristic' insider will trade more aggressively than the rational one– to result in a higher probability to earn more profits (Liu, Qi, & Li, 2019).

15.2.2 AVAILABILITY HEURISTIC

As scenarios and instances become available, they get frequently employed. More so when people start assessing their plausibility and frequency of occurrence, aligned to specific events (Tversky & Kahneman, 1974).

15.2.2.1 REAL LIFE EXAMPLES

- When Medical practitioners diagnose a patient's condition, they are most likely to arrive at a similar diagnosis – like the condition the patient had encountered in the recent past (Poses & Anthony, 1991).
- Only after experiencing a natural disaster, people rush in to purchase disaster insurance (Karlsson, Loewenstein, & Ariely, 2008).
- The teaching course which asks for more critical comments from the students is seemingly rated better, during course evaluations (Fox, 2006).

15.2.3 AFFECT HEURISTIC

Feelings that hinge upon the specific measures of being good or bad experience a state of consciousness depending on the quality of the stimulus – positive or negative (Slovic et al., 2007).

15.2.3.1 REAL LIFE EXAMPLES

- Emotions drove consumers at Las Vegas hotels differentiate their preference, giving equal weightage to the parameters of price and service (Ro et al., 2013).
- Watching pictures of flooded homes make houseowners consider flooding a considerable risk. For an effective risk communication, affect is a critical parameter (Keller et al., 2006).
- Consumers who encounter interacting with innovative product develop a biasness basis their positive or negative feelings (King & Slovic, 2014).
- The key to positioning the product in the hospitality industry is to understand what the customer feels, i.e., the customers perception about the brand (Ro, Lee, & Mattila, 2013).

15.2.4 PEAK AND EFFECT HEURISTIC

More than the objectivity measure, the consumer's evaluation of an experience is shaped by its most intense and the final moments (Cockburn et al., 2015).

15.2.4.1 REAL LIFE EXAMPLES

- While in labor, delivery is the factor influencing the entire motherhood experience (Chajut et al., 2014).
- The amount of prevalent negativity in the workplace is a good predictor of an employee quitting the job (Clark & Georgellis, 2006).
- Peak experiences and positive outcomes have the highest customer satisfaction especially in service industries (Verhoef et al., 2004).
- An individual's affective and hedonic experiences have a strong lineage to the peak-end effects (Cockburn, Quinn, & Gutwin, 2015).
- One-Time Wonders and Gamblers can evaluate their experiences better than consistent winners and small-time wonders (Yu et al., 2008).

15.2.5 SOCIAL PROOF/REFERENCE HEURISTIC

"Using the actions of others to infer the value of a course of action," (Rao et al., 2001).

15.2.5.1 REAL LIFE EXAMPLES

- Good reviews and better ratings drive the demand in online sales (Amblee & Bui, 2014).
- People tend to donate charitable organizations when they find others donating in them (Shearman & Yoo, 2007).
- The signages in supermarkets are influenced by consumer choices (Salmon et al., 2015).
- The response from guests is most positive when savings are for charity. However, when the same guests participate for a social proof, the impact of savings gets reduced (Shang, Basil, & Wymer, 2010).

15.2.6 ILLUSION OF ATTENTION HEURISTIC

Inattentional blindness and deafness demonstrate 'the illusion of attention' (Simons & Chabris, 1999).

This category points out that, an individual's focused attention may make the individual oblivious to notice something which is otherwise obvious.

15.2.6.1 REAL LIFE EXAMPLES

- Firmly entrenched in the cozy comforts of the train coach, one is deeply engrossed in the bestseller of one's favorite author. With all the cacophony around, there's also an announcement by the loco driver that the journey's been a bit delayed. But all go unnoticed. Completely oblivious to the surroundings, the shrill whistle of the engine approaching the destination also goes unnoticed. Only when the last line of the paragraph is about to end, we wake up from the slumber only to see the coach pulling out of our alighting station!
- In 'The Invisible Gorilla' (Chabris & Simons, 2011) examined the false impressions to minimize the negative impacts. In the pathbreaking study, volunteers watched a 1-minute clipping of a basketball game and were to count the number of passes being made. Quaintly enough, just at the 30-sec mark, a woman covered from head-to-toe and dressed in a gorilla outfit stealthily moved in front of the screen, stared at the camera, beat her chest – all gorilla style and strolled back, casually. When asked, majority of the volunteers answered that they never saw any movement in the room.

The Cycle safety campaign of the city of London drew the motorists' attention to the perils of inattentional blindness and challenges of inattentional deafness (MacDonald & Lavie, 2011) (Figure 15.2).

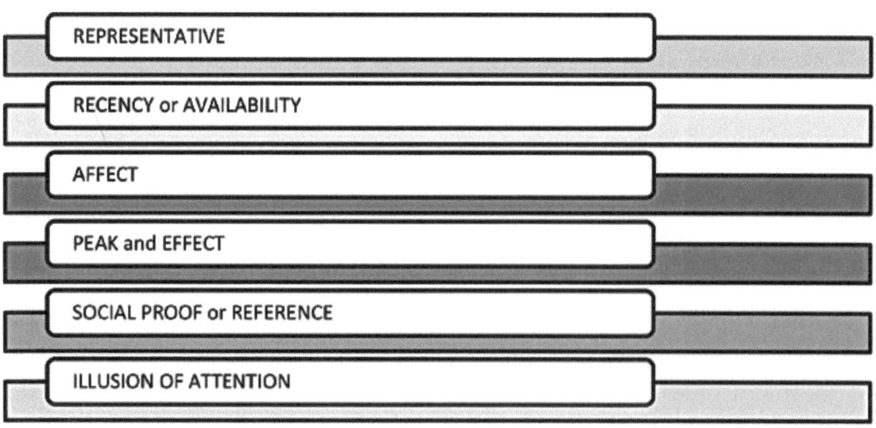

FIGURE 15.2 The heuristics.

15.3 THE BIASES

According to research, consumers are behaviorally neither economic nor rational. Such behavioral biases come in the way of consumers taking responsibility for their deeds of action. It defies logic for business houses to try and exploit these consumer biases rather than come out with options to respond to the behaviors.

15.3.1 ANCHORING BIAS

Most of the time, individuals rely on their initial (intuitive) value to make estimates. This value becomes the focal point for identifying and formulating the problem statement. With passage of time, biasness (towards the initial value) starts creeping in. This is the 'anchoring bias' (Tversky & Kahneman, 1974).

15.3.1.1 REAL LIFE EXAMPLES

- The decision-making process has major ramifications from the anchoring effect (Furnham & Boo, 2011).
- The aggressive first mover is normally the driving party in negotiated outcomes. In competitive markets like Real-Estate, the final settlement price is a build-on from the ask price (Aycock, 2000).

15.3.2 OMISSION BIAS

Despite containing the same (negative) consequence, a commit-to-action decision is more negatively evaluated than an omit-to-action one (Kordes-de Vaal, 1996).

15.3.2.1 REAL LIFE EXAMPLES

- Parents often demonstrate their reluctance to vaccinate children (with a new drug), more so when the after-effects are not known (Baron & Ritov, 1990).
- Being susceptible to the omission bias, there's an uptick in marketing and action-enabled sales training programs (Crego & Heuvel, 2012).
- An observable convention in the stock market – firms having violation by commission have a more severe run down when compared to firms having violations by omission (Wiles et al., 2010).

- Jet pilots with a higher sense of personal accountability commit less errors than pilots with lower sense of accountability, who by relying more on automated systems are error-prone (Mosier et al., 1998).

15.3.3 NEGATIVITY BIAS

As a convention, positive information when compared to negative information has a weaker influence on evolutionary outcomes.

15.3.3.1 REAL LIFE EXAMPLES

- More people believe the information to be true, when the information is veered towards the negative side (Hilbig, 2009).
- In any qualitative employee survey questionnaire, it is common to find negative comments from the dissatisfied employees when compared to their satisfied colleagues (Poncheri, 2007).
- Negative Framing or the Fear of Losing incentivizes the average consumer much more than positive framing or the chance of gaining (Dhar & Goldsmith, 2013).
- The difference in the Negativity Bias stems from a better categorization of positive biases (Kiken & Shook, 2011).

15.3.4 CONFIRMATION BIAS

Seeking or interpreting available evidence to buttress available beliefs, expectations or hypothesis is confirmation bias.

15.3.4.1 REAL LIFE EXAMPLES

- In the year 1975, Stanford University carried out research on Suicide. Handing out a pair of 'suicide note' to each participant – one note from a random individual and the other from a person attempting to take his/her own life. Once all the notes came back duly responded, students in the undergrad class were asked to distinguish between the genuine notes and the faked ones.

  Though the setup was a put-on, the study revealed that though 50% of the notes were genuine, the scores remained fictitious.

  With deception revealed, the researchers were directed to the main reason for research – to gage if the responses to thinking were right or wrong. This too a deception.

Adopting to reasoning and strategically aligned, the evolved humans of today are both hypersocial and niche (Mercier & Sperber, 2017).

- Science tells us something different from what we personally tell ourselves (Gorman & Gorman, 2017). Rudimentary beliefs may not just be false, but also potentially harmful. A case in point being the growing w-o-m murmur that vaccines are hazardous to health. Time and again science has proven that rather being hazardous, vaccination has been one of the triumphs of modern medicine (Gorman & Gorman, 2017).

Despite strong scientific proof of the positives emanating from vaccination, the anti-vaxxers remain stationary. A classic example coming from the past occupant of the White House in USA. President Donald Trump was proud to proclaim that due to this hesitancy, although he had to get his son vaccinated, he never followed the timetable as recommended by pediatricians.

15.3.5 SIMPLICITY BIAS

Hence, despite apparent complexity (of the world around us), there is an inherent bias towards simplicity. This holds true for all activities in everyday life.

15.3.5.1 REAL LIFE EXAMPLES

1. **Netflix:** Be it work or at leisure, office or at home, one's constant companion today is the internet. Netflix is an internet streaming platform offering TV shows, documentaries, animation, movies, and many more. What makes the Netflix experience more engrossing is that, using algorithms that track the viewing pattern of individuals, the platform succeeds in providing a hassle-free viewing experience.
2. **Google:** Though being launched over two decades back, Google has remained steadfast in its mission of organizing the world's information and making it universally accessible. Even though the information technology (IT) sector has encountered many an inclement and rough weather, the universal appeal that Google has, makes the brand soldier on.

   Though being the basic tenet of customer experience, the company's outdated rules or procedures often make simplicity to be ruled out-of-favor (Gingiss, 2020).

15.3.6 ACTIVITY BIAS

Actions that can be termed detrimental or which appear outrightly silly can be attributed to Activity Bias.

15.3.6.1 REAL LIFE EXAMPLES

One of the prominent examples of Activity Bias is observing the behavior of passengers boarding the airplane. Every time there's an announcement of boarding, there invariably is at least one fellow passenger who will do whatever it takes – even break the line, to be the first person to board the airplane – as if, boarding the aircraft first, will enable him/her to reach the destination faster than the others.

Also, when we consider the returns associated with the stock markets, prudent investors are generally guided by the age-old adage of 'buy low sell high.' However, there are also instances of amateur investors, who driven by and unable to control their emotions do just the opposite.

15.3.7 GRATIFICATION BIAS

Gratification Bias can be understood as that specific component of the unconscious mind which is purely driven by the basic instincts. The concept of Gratification Bias was first used by Sigmund Freud when he described the role of the 'id' (Good Therapy, 2015).

15.3.7.1 REAL LIFE EXAMPLES

Pertinent real-world examples would be:

- To not be able to stop the desire to go in for purchasing a new entrant in the market (especially if it is a luxury commodity and hence elastic in demand), even by spending a fortune – taking a high-interest loan. The consumer could easily have deferred the said purchase till s/he had saved enough money to buy without the burden of taking the loan.
- To be giving in to the temptation of partying with friends on a work-night, instead of catching up on good night sleep.

15.3.8 EMOTIONAL BIAS

This type of biases most typically arises during the process of Decision Making, where there is a lot of spontaneity based on the consumers' personal likes and dislikes, and even feelings.

15.3.8.1 REAL LIFE EXAMPLES

1. **Loss-Aversion Bias:** This is a Bias related to the Investors or their Portfolio Managers. For ex: There is a stock, though blue-chip, which is currently so down in the market that the thought of even selling is ruled out.

   While selling the stock would have at least given some liquidity and purchasing power to scout for other higher-quality stocks, for reasons best known, the concerned Investors/Portfolio Managers do not want to admit the loss (notionally) with the hope that the stock (it being a blue-chip one) will certainly bounce back, one day.

2. **Overconfidence Bias:** "I have an edge that you (and others) do not." This is the basic principle that drive the concept of overconfidence bias.

3. **Endowment Bias:** Doing something on our own is more precious than not doing something. In Financial markets, if as an investor we look at the 'losing' stock, most of the other stocks may/will turn profitable with signs of recovery, but we still will hold on to this 'losing' stock – only because we believe this stock is the best in its sector! (Figure 15.3).

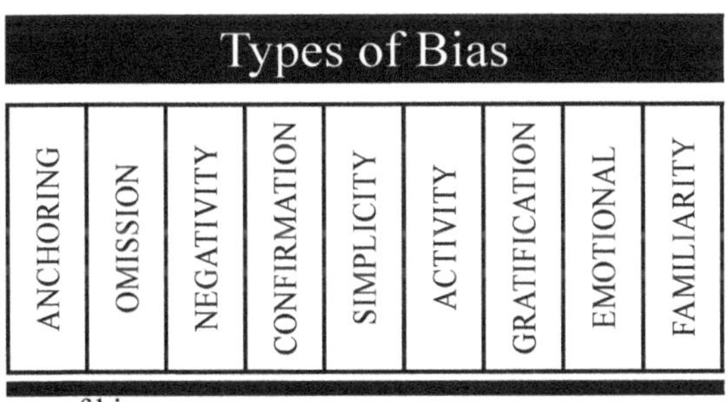

FIGURE 15.3 Types of bias.

15.3.9 FAMILIARITY BIAS

Individuals who dislike ambiguity, tend to be comfortable with the familiar and look for ways to avoid the unknown bred the concept of Familiarity Bias.

An experiment by 92 (four-person chains) individuals to unearth the fitment and survival of positive over negative story events (albeit unambiguously) was carried out (Bebbington, MacLeod, Ellison, & Fay, 2017).

15.4 SUMMARY AND CONCLUSION

Frictions in Decision Making can be reduced by applying Cognitive Bias. Such applications of Heuristic principles enable convert complex decisions into simpler ones (Pendergrass, 2019).

According to prominent social scientists, each decision that a consumer makes is predictable, yet irrational (Thaler, Kahneman, Ariely, & Cialdini, 2012).

Consumers take recourse to mental shortcuts when they are not concentrating on the problem (Kahneman, 2012). Kahneman names this style of thinking as the System 1 style of thinking. Consumers use this style of thinking to arrive at a decision.

System 1 style of thinking relies on estimates and approximations. Such approximations are 'biases' or 'heuristics' and can lead the customers/ consumers to behave in a predictable yet irrational way (Figure 15.4).

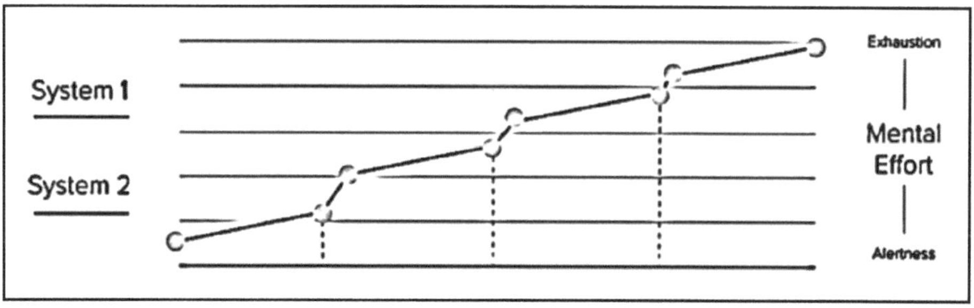

FIGURE 15.4 Consumer thinking styles.

Source: Adapted from Jochen 2021, https://www.convertize.com/consumer-behavior/

In his book, Predictably Irrational, Dan Ariely takes this argument further by identifying and isolating the most common biases that seem to occur in our day-to-day lives.

Throwing caution to the wind, the behaviorist in Ariely argues that the consumer behavior is not to measure the cost vs benefit accruals. On the contrary, Ariely argues that consumer decisions are always based on guess-estimations and intuitions.

Intelligent marketers identify such cognitive effects to draw the attention of their customers, and thereby avoid falling prey to the biases.

KEYWORDS

- **behaviorism**
- **cognitive bias**
- **consumerism**
- **decision making**
- **heuristics**

REFERENCES

Amblee, N., & Bui, T., (2011). Harnessing the influence of social proof in online shopping: The effect of electronic word of mouth on sales of digital microproducts. *International Journal of Electronic Commerce*, *16*(2), 91–114.

Ariely, D., (2009). *Predictably Irrational: The Hidden Forces that Shape our Decisions*. Revised and expanded edtions. Harper Collins.

Bebbington, K., MacLeod, C., Ellison, T. M., & Fay, N., (2017). The sky is falling: Evidence of a negativity bias in the social transmission of information. *Evolution and Human Behavior*, *38*(1), 92–101.

Chabris, C., & Simons, D., (2010). *The Invisible Gorilla: And Other Ways Our Intuitions Deceive Us*. United States and Canada Harper Collins.

Chajut, E., Caspi, A., Chen, R., Hod, M., & Ariely, D., (2014). In pain thou shalt bring forth children: The peak-and-end rule in recall of labor pain. *Psychological Science*, *25*(12), 2266–2271.

Chen, Y., Tso-Jen, C., & Lin, C., (2016). The analyses of purchasing decisions and brand loyalty for Smartphone consumers. *Open Journal of Social Sciences*, *4*(7), 108–116.

Cockburn, A., Quinn, P., & Gutwin, C., (2015). Examining the peak-end effects of subjective experience. In: *Proceedings of the 33rd Annual ACM Conference on Human Factors in Computing Systems* (pp. 357–366). ACM DL. New York NY.

Furnham, A., & Boo, H. C., (2011). A literature review of the anchoring effect. *The Journal of Socioeconomics*, *40*(1), 35–42.

Goldsmith, K., & Dhar, R., (2013). Negativity bias and task motivation: Testing the effectiveness of positively versus negatively framed incentives. *Journal of Experimental Psychology: Applied*, *19*(4), 358.

Gorman, S., & Gorman, J., (2017). *Denying to the Grave – Why We Ignore the Facts That Will Save Us* (1st edn.). Oxford University Press, USA.

Gutwin, C., Rooke, C., Cockburn, A., Mandryk, R. L., & Lafreniere, B., (2016). Peak-end effects on player experience in casual games. In: *Proceedings of the 2016 CHI Conference on Human Factors in Computing Systems ACM DL* (pp. 5608–5619). New York NY.

Kahneman, D., (2011). *Thinking Fast and Slow*. Farrar, Straus, and Giroux.

Kiken, L. G., & Shook, N. J., (2011). Looking up: Mindfulness increases positive judgments and reduces negativity bias. *Social Psychological and Personality Science*, *2*(4), 425–431.

King, J., & Slovic, P., (2014). The affect heuristic in early judgments of product innovations. *Journal of Consumer Behavior*, *13*(6), 411–428.

Krawczyk, M. W., & Rachubik, J., (2019). The representativeness heuristic and the choice of lottery tickets: A field experiment. *Judgment and Decision Making*, *14*(1), 51.

Liu, H., Qi, L., & Li, Z., (2019). Insider trading, representativeness heuristic insider, and market regulation. *The North American Journal of Economics and Finance*, *47*, 48–64.

MacDonald, J. S. P., & Lavie, N., (2011). Visual perceptual load induces inattentional deafness. *Attention, Perception, and Psychophysics*, *73*(6), 1780–1789.

Mercier, H., & Sperber, D., (2017). *The Enigma of Reason – A New Theory of Human Understanding.* Harvard University Press.

Ro, H., Lee, S., & Mattila, A. S., (2013). An affective image positioning of Las Vegas hotels. *Journal of Quality Assurance in Hospitality & Tourism*, *14*(3), 201–217.

Salmon, S. J., De Vet, E., Adriaanse, M. A., Fennis, B. M., Veltkamp, M., & De Ridder, D. T., (2015). Social proof in the supermarket: Promoting healthy choices under low self-control conditions. *Food Quality and Preference*, *45*, 113–120.

Shang, J., Basil, D. Z., & Wymer, W., (2010). Using social marketing to enhance hotel reuse programs. *Journal of Business Research*, *63*(2), 166–172.

Simons, D. J., & Chabris, C. F., (1999). Gorillas in our midst: Sustained inattentional blindness for dynamic events. *Perception, 28*(9), 1059–1074.

Sprinkle, Z. J., (2019). *Heuristics in Construction Project Management.* Masters Thesis, Virginia Tech.

Van, D. H. C., Alison, L., & Crego, J., (2012). How uncertainty and accountability can derail strategic 'save life' decisions in counter-terrorism simulations: A descriptive model of choice deferral and omission bias. *Journal of Behavioral Decision Making*, *25*(2), 165–187.

Wiles, M. A., Jain, S. P., Mishra, S., & Lindsey, C., (2010). Stock market response to regulatory reports of deceptive advertising: The moderating effect of omission bias and firm reputation. *Marketing Science*, *29*(5), 828–845.

CHAPTER 16

Impact of Consumers' Knowledge and Use of Food Labels on Willingness to Pay Using SmartPLS

MEHA SAXENA,[1] ANKITA JAIN,[2] and M. K. SHARMA[3]

[1]*Sr. Assistant Professor, School of Management, IIS (deemed to be University), Jaipur*

[2]*Professor, School of Management, IIS (deemed to be University), Jaipur*

[3]*Dean, School of Commerce & Management, IIS (deemed to be University), Jaipur*

ABSTRACT

With the increase of food scares, the regulatory bodies, manufacturers, and researchers are analyzing the provision of food label information (FLI) on packaged foods for an informed food choice (Hagmann & Siegrist, 2020). Various studies worldwide have assessed the usage pattern of food labels by consumers and reported it mostly to be consistent (Campos et al., 2011). However, the self-reported usage of food labels is found to be high, consumers sometimes face difficulty in understanding the technicalities of the information provided. Also, packaging is a very cost-intensive activity. Thus, as an incentive to bear this extra cost, manufacturers expect higher price returns on their product. But is the consumer willing to pay more for quality assurance and proper information? This chapter aims at analyzing the usage pattern of consumers and their knowledge about food labels in five major cities of Rajasthan, India. Also, it tries to develop a relationship between the use and knowledge of consumers about food labels with their willingness to pay (WTP) an extra price for proper FLI using SmartPLS. It was found

Sustainability in Marketing Practice: Strategies for Industry 4.0.
Subrata Chattopadhyay, Udit Chawla, (Eds.)

that consumers are using food labels for making purchase decisions, but their knowledge remains limited in the field. They are willing to pay 10–15% more price for proper information. Though use has a positive moderately strong effect on WTP, Knowledge has a mild negative effect on consumers' WTP.

16.1 INTRODUCTION

Food has always occupied a major consideration in day-to-day decision making of a household. But there has been a shift in the overall food consumption pattern. Earlier, there was a reliance on locally available traditional food products. Fresh produce was more popular leading to limited availability of seasonal foods only. But, with the developments in the field of Food Technology and globalization of markets, supermarket shelves are loaded with a wide variety of packaged food products. The reason lies in the increasing needs of customers for reduced cooking time, convenience, variety, and globalization of food cuisines, the food market is shifting towards processed and packaged food (Choicequality et al., n.d.).

But this development comes in hand with a lot of food scares about food quality and safety for consumption. Customers are extremely sensitive about the quality of food products that they consume in order to ensure that it is safe to consume and as good as home-cooked food (Grunert, 2005). Apart from the probability of the presence of Adulterants and Allergens in the food, people have also become very much conscious about the health aspect of the food they consume (Krystallis & Ness, 2005).

With the advent of so many food scares among consumers, food safety issues are attracting a lot of public attention, which has led to increasing stress on the development of food policies. Consumers have become more conscious about how the food is produced and are very critical about quality differentiation of food products. All these considerations have led to an increased focus on Food Quality and Safety (Van Rijswijk & Frewer, 2008).

The customer food quality perception has a direct impact on his WTP for a food product and thus has a direct impact on effective demand for food products in the market (Miškolci, 2011).

Different researchers have developed different approaches to study consumers' food quality perception, such as the means-end approach, expectancy-value approach, economics for information approach, satisfaction/dissatisfaction approach, and total food quality approach.

Among these, the most widely used model is the total food quality (TFQ) model, which was proposed by Grunert in 1996 (Grunert, 2005). According to this model, quality cues can be intrinsic or extrinsic. While the intrinsic cues relate to the product characteristics, the extrinsic cues can be attributed to everything else. Some major extrinsic cues can be considered as brand, product origin, and food labels (Latvala, 2004). Thus, food label information (FLI) has become very critical and has gained a lot of importance in the packaged food industry. The consumers' demand for proper FLI has especially increased after the advent of COVID-19. People are more inclined to buy packaged food products as a precautionary measure. They feel that packaged food is both hygienic and a limited contact product. Different countries, including India, have made it mandatory for manufacturers to provide proper FLI so that consumers can make an informed food choice (Vemula et al., 2013).

People are even willing to pay an extra price to ensure that the product they buy is properly packaged and safe to consume (Voon et al., 2011). They expect that the food they consume is safe and healthy. They wish to make an informed food choice. They rely heavily on FLI for making a food choice. But willingness to pay (WTP) is affected by consumers' knowledge about the nutrition information (Chiappa, 2015).

Knowledge about food labels may include factors like whether consumers have basic knowledge about terminologies like "use by" date, "expiry" date, "best before" date, etc., or calorie counts and other nutrients. Are they aware of packaging regulations like mandatory information and non-mandatory information, etc. (Nasreddine et al., 2014).

But it has been observed that many consumers are not able to understand the information provided on food labels due to their limited knowledge and the technicality of the terms used (Vemula et al., 2013). Also, the extent of use of FLI varies among consumers based on their demographic profile, such as age, gender, qualification, etc., and situational factors like the type of product, time, and place of purchase, etc. (Vijaykumar et al., 2013).

The present study focuses on examining the perceived knowledge and the actual level of knowledge about FLI and the gap therein. The second aim of the chapter is to find the usage pattern among consumers. The final aim of the study is to analyze the impact of knowledge and usage pattern of consumers on their WTP a premium price in order to ensure quality food products.

16.2 CONCEPTUAL MODEL

The conceptual model of the study is given in Figure 16.1. The model comprises consumers' knowledge about FLI and their use of food labels as independent variables which in turn lead to their WTP as the dependent variable.

From previous studies, it has been observed that if a consumer has some knowledge about the importance of food labels, they are willing to pay more for getting this information on food labels to make an informed food choice. Also, when this knowledge gets converted into use (Miller & Cassady, 2015), we can say that those consumers who are using food labels are more likely to pay extra for getting this information in the desired format (Darkwa, 2016). Based on these assumptions, the model was framed accordingly and the impact of consumers' knowledge and use of food labels on their WTP was analyzed.

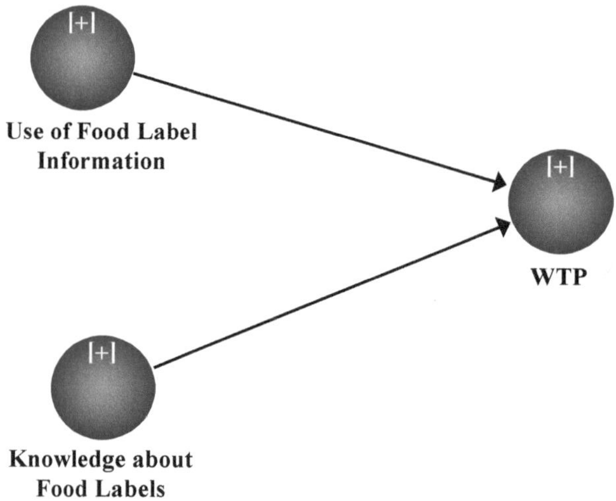

FIGURE 16.1 Conceptual model.

16.2.1 *OBJECTIVES OF STUDY*

- To study consumers' knowledge about FLI;
- To observe the usage pattern of FLI;
- To determine the impact of consumers' knowledge and use of FLI on their WTP for packaged food products;
- To assess the global predictability of the results.

16.2.2 *HYPOTHESES OF STUDY*

- **Hypothesis 1:**
- **Ha$_1$:** There is a significant variation in consumers' usage pattern of food labels demographically.
- **Ha$_2$:** There is a significant variation in consumers' knowledge about FLI demographically.

- **Ha_3:** There is a significant association between factors consumers' knowledge and usage pattern of food labels and their WTP.

16.3 METHODOLOGY

16.3.1 STUDY DESIGN

A primary study was conducted in a cross-sectional Urban Consumer market using Stratified random sampling in the month of March–July 2021 in 5 cities of Rajasthan, India viz. Jaipur, Jodhpur, Udaipur, Ajmer, and Kota. Each city was further divided in 4 zones.

16.3.2 SAMPLE CHARACTERISTICS

Respondents were drawn from different demographic profiles for the study. Only adult respondents who were majorly responsible for decision regarding grocery shopping for the family were taken into consideration. According to the Morgan formula for infinite population, at 5% level of significance and 5% margin of error, a sample size of 384 is recommended. A total of 400 respondents participated in the study. But, after removal of outliers, a total of 392 records were used for analysis.

16.3.3 SCALE DEVELOPMENT

In order to record the consumers' knowledge about FLI, items like their understanding of difference between the expiry date, the use by date and the best before date, the order of display of ingredients, the mandatory and non-mandatory components of nutritional information panel, difference between energy and calories, measurement in volume of one tablespoon and their idea about processed and preserved foods (Joel et al., 2016). For Use of food labels, respondents were asked questions related to the frequency of use of labels and situations in which they especially make use of food labels for making purchase decisions (Cowburn & Stockley, 2014). Their WTP (WTP) extra for proper display of FLI was checked on a 5-point Likert scale for different food categories (Chiappa, 2015). For Knowledge and WTP construct, Agreement type Likert Scale was used, while Frequency Likert scale was utilized for use construct (Table 16.1).

TABLE 16.1 Scale Development

Construct	Item	Description
Knowledge about food labels	K1	Is it compulsory for all producers to put nutrition Information panel on all food products?
	K2	Are expiry date, Use by date and best before date synonymous?
	K3	Ingredients are displayed in ascending order of their quantity in formulation.
	K4	One tablespoon is always equal to 7 ml liquid.
	K5	Processed foods are unhealthy.
	K6	Preserved foods are unhealthy.
Use of FLI	U1	Comparison with similar products.
	U2	To check ingredients, allergens.
	U3	To check Mfg/expiry date.
	U4	To check compliance with food quality standards.
	U5	To check nutrition panel or part of it.
	U6	To check directions of use.
WTP	WTP1	Dairy products.
	WTP2	Fruit and vegetable products.
	WTP3	Staples.
	WTP4	Bakery and confectionary.
	WTP5	Drinks and beverages.

16.3.4 INSTRUMENT DEVELOPMENT

As the study was based on primary data, hence a questionnaire was designed and administered to the respondents by means of exit poll outside supermarkets in different zones for study. The questionnaire was broadly divided into three parts. The first part of the questionnaire aimed at demographic profiling of the respondents. The second part of the questionnaire covered the knowledge about the FLI and the usage pattern of the consumers for food labels while making purchase decision. The last part of the questionnaire aimed at knowing the extent of consumers' WTP a premium price for prepackaged foods with proper FLI as an indication of good quality products. The questions in the second and the third part of the questionnaire were designed on a five-point Likert scale.

16.3.5 STATISTICAL ANALYSIS

The data collected by means of questionnaire was first entered in MS Excel 2007. After performing filtering and coding, the data was exported to SmartPLS 3.0 for further statistical analysis.

16.4 RESULTS

16.4.1 *VALIDATION OF MEASUREMENT MODEL*

In order to validate the relationship between the constructs and their response items, a measurement model was constructed having two levels: the outer reflective model and the inner formative model (Aziz & Kamaludin, 2014). For the validation of the model, the data was analyzed in SmartPLS 3.0.

16.4.1.1 OUTER REFLECTIVE MODEL

Selection of the items for each construct was based on previous studies in the field. As the first step towards valuation of the reflective model, the Cronbach Alpha value of each construct was checked and found to be above 0.7.

After checking the Cronbach Alpha values, the next level of validation was done to check the outer loading values of the items and convergent validity of the model to check the agreement between the multiple items of a latent construct. Convergent validity was checked using two measures: average variance extracted (AVE) having a value more than 0.5 and composite reliability (CR) within the range 0.7 to 0.9 (Hair et al., 2019).

The Measurement model and the values of the outer loadings for the constructs under study are given in Figure 16.2.

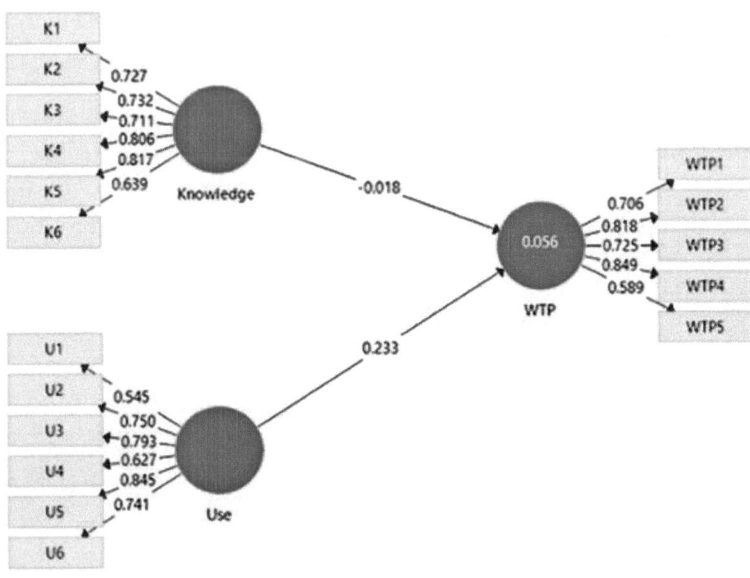

FIGURE 16.2 Measurement model.

Though the items K6, U1, U4, and WTP5 did not cross the minimum requirement of outer loading, i.e., 0.708, they were still retained in the study as the AVE of their constructs was more than 0.5. The CR of the constructors also well within the range. The values of the outer loadings and convergent validity for the constructs under study are given in Table 16.2.

TABLE 16.2 Convergent Validity

Construct	Item	Loading Weight	Cronbach's Alpha	CR	AVE
Knowledge about food labels	K1	0.727	0.863	0.879	0.550
	K2	0.732			
	K3	0.711			
	K4	0.806			
	K5	0.817			
	K6	0.639			
Use of FLI	U1	0.545	0.816	0.866	0.524
	U2	0.750			
	U3	0.793			
	U4	0.627			
	U5	0.845			
	U6	0.741			
WTP	WTP1	0.706	0.792	0.859	0.552
	WTP2	0.818			
	WTP3	0.725			
	WTP4	0.849			
	WTP5	0.589			

After validating the internal consistency of all the constructs, next the uniqueness of each construct was checked with the help of Discriminant Validity. The litmus test for Discriminant Validity was chosen as the Heterotrait Monotrait ratio (HTMT) which should be less than 0.85 (Henseler et al., 2015). The HTMT values of all the three constructs of study are presented in Table 16.3.

TABLE 16.3 Discriminant Validity Using HTMT Ratios

	Knowledge	Use	WTP
Knowledge	–	–	–
Use	0.178	–	–
WTP	0.087	0.282	–

From Table 16.3, it was evident that all the constructs were distinct and unique. After validation of the outer reflective model text validation of the inner formative model was done.

16.4.1.2 FORMATIVE MODEL

To validate the formative model, model proposed by Hair et al. (2019) was utilized. For this, first the Convergent Validity of the constructs was checked using the Redundancy Analysis. For this, a global construct was created using single item at the time of research design for the dependent variable that is WTP and this was correlated with the other constructs of the Formative Model. As the path coefficient was found to be 0.801 the value was considered satisfactory and convergent validity was achieved.

For the next stage of validation, the multi-collinearity of the indicators had to be verified as in case of collinearity, it may affect the structural model of the study. For this purpose, the values of VIF should be less than equal to 3 (Henseler et al., 2015). The VIF values of all the items is tabulated in Table 16.4 and it was found that all the values within the range thereby eliminating any chances of multi-collinearity between the item constructs.

TABLE 16.4 Multi-Collinearity Analysis by VIF Values

Indicator	VIF
K1	2.013
K2	1.483
K3	2.136
K4	1.810
K5	2.175
K6	2.154
U1	1.258
U2	1.661
U3	1.965
U4	1.565
U5	2.375
U6	1.852
WTP1	1.435
WTP2	1.962
WTP3	1.519
WTP4	2.080
WTP5	1.300

For the last stage of validation of the formatted model the importance and relevance of the outer weights was checked using bootstrapping with 10,000 iterations. The beta estimates and the t-statistic values of each construct was checked at p values less than 0.05. All the constructs were considered significant. The results of bootstrapping are summarized in Table 16.5.

TABLE 16.5 Importance and Relevance of Outer Weights

Hypothesis	Relationship	Beta	t-Value	p-Value	Decision
H1	Knowledge → WTP	–0.018	0.616	0.0003	Supported
H2	Use → WTP	0.233	3.79	0.0000	Supported

As p values were less than 0.05, both H1 and H2 were supported (Hair et al., 2019). As the measurement model was validated at both the reflective and formative levels, next the hypothesis regarding the relationship between the latent constructs had to be tested for the structural model.

16.4.2 VALIDATION OF STRUCTURAL MODEL

The structural model of the study is formative as both the knowledge about the food labels and the usage pattern affect the WTP extra for the proper FLI.

To test the structural model, first r squared values were analyzed. The R^2 value of WTP was 25.67%. The effect size of each relationship was checked using the f square values which is summarized in Table 16.6.

TABLE 16.6 Effect Size Using f^2 Value

Construct	f^2 Value
Knowledge	0.056
Use	0.1721

From the table is it evident that the effect size of the relationship between the use of food labels and the WTP was moderate (17.21%), but knowledge about food labels had a low effect size (5.6%) on the WTP for FLI (Shmueli et al., 2019).

The multi-collinearity of the model was once again checked using the VIF and the values are summarized in Table 16.7.

TABLE 16.7 Multi-Collinearity Using VIF Values

Construct	Inner VIF Values
Knowledge	1.037
Use	1.036

As both the values were less than 3, it was concluded that there is no multi-collinearity in the inner model as well. The model fit of the study was done using the SRMR value which was less than 0.8.

Finally, PLS Predict was run for the model to check the global predictability of the model. As the items of WTP were found to be normally distributed, the RMSE values of the PLS model and Linear model were compared. As Q^2 value of all the items was positive and in majority of the items, the residual was found to be positive, indicating high predictability of the model. Thus, this model can be referred for further studies globally (Shmueli et al., 2019). The summary of PLS predict results are given in Table 16.8.

TABLE 16.8 Residuals by PLS Predict

	Q^2 Predict	Residual
WTP1	0.013	0.023
WTP2	0.009	0.018
WTP3	0.018	0.107
WTP4	0.007	–0.07
WTP5	0.01	0.022

16.5 CONCLUSION

From the above analysis, it has been observed that the self-proclaimed usage of FLI has gone up in the last decade. These results are commensurate with many studies conducted globally as well (Campos et al., 2011). Consumers have started realizing the importance of Food labels and prefer making an informed choice to ensure food safety and quality (Krystallis & Ness, 2005). Consumers from all age groups, genders, levels of qualification and income groups show their interest in FLI. Also, today, all manufacturers are compelled to provide proper information in their products. But the gap lies in the desire to utilize and actual efficient utilization of this information (Hagmann & Siegrist, 2020). Though most of the consumers are willing to use the FLI, their usage is limited due to lack of knowledge about the information provided on the food labels. People are not clear about the mandates regarding nutrition information. They are not able to process the nutritional information panel as well. Though they claim to make purchase decision based on food labels, they are not able to understand what food is good food based on the information provided on the labels (Saxena et al., 2021). Though FSSAI is making a lot of efforts in different dimensions in order to improve consumers' awareness and knowledge about food labels, more efforts need to be made by companies and researchers in order to educate the customers about the importance of food labels and how to use them in an efficient way to make a smart food choice.

Also, it has been observed that the knowledge has a low negative significant and usage pattern of food labels has a positive and significant impact on consumers' WTP for superior food quality. They are ready to pay 10–15% price for proper display of FLI so that they can refer to the same while making suitable and healthy food choice.

16.6 MANAGERIAL IMPLICATIONS

From the importance – performance map given in Figure 16.3, it is clear that as a construct, knowledge about food labels has a very low importance in the determination of WTP. Though performance is not very low. This shows that if regulatory bodies make some efforts in improving the knowledge of the consumers, it may increase people's WTP for proper food labels. But, in all, great results may not be achieved. On the other hand, the use of food labels is already in the high performance–high importance quadrant, showing that any efforts made by manufacturers and regulatory bodies to encourage the use of food labels may bring in a positive attitude of consumers towards paying a premium price for proper dissemination of FLI. If there is an increase in the use of FLI by 1 unit from 80.333 to 81.333, WTP will rise from 74.874 to 75.131 (Hair et al., 2019).

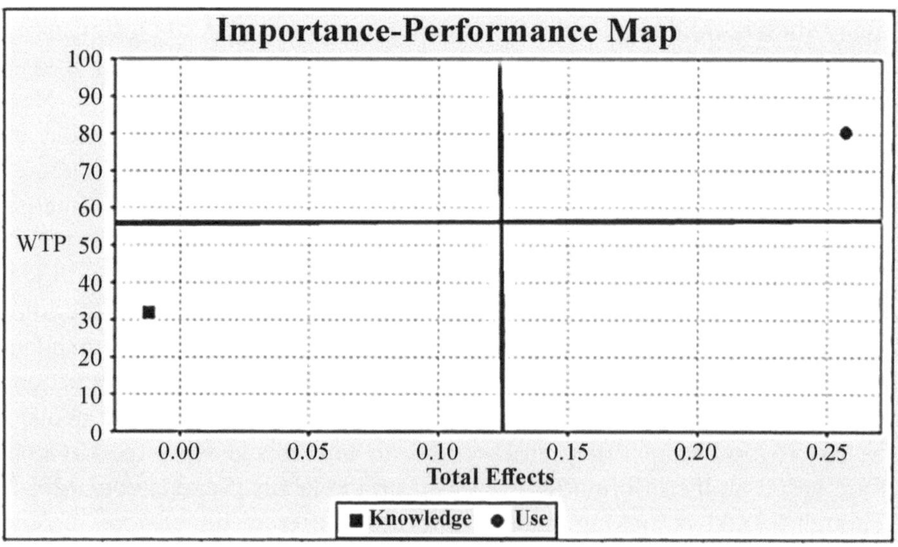

FIGURE 16.3 Construct wise IPMA for knowledge and use.

On further analysis of importance-performance of items of constructs, as shown in Figure 16.4, none of the items pertaining to knowledge construct performed well in contributing towards WTP. The reason is quite obvious. As the knowledge level is very low compared to the minimum desired level, it is not able to contribute at all in WTP. This shows that a lot of efforts need to be put in order to improve the knowledge about food labels in masses so that they may understand its importance and make wise food purchase decisions. But all the items of usage pattern show a strikingly opposite behavior. All the items lie in the high importance high performance quadrant. Though, item U4 related to checking of Quality standards shows lesser importance for the consumers and U1 related to comparison of similar products also shows comparatively less importance, all other items have high importance and performance in WTP a higher price for proper display of information.

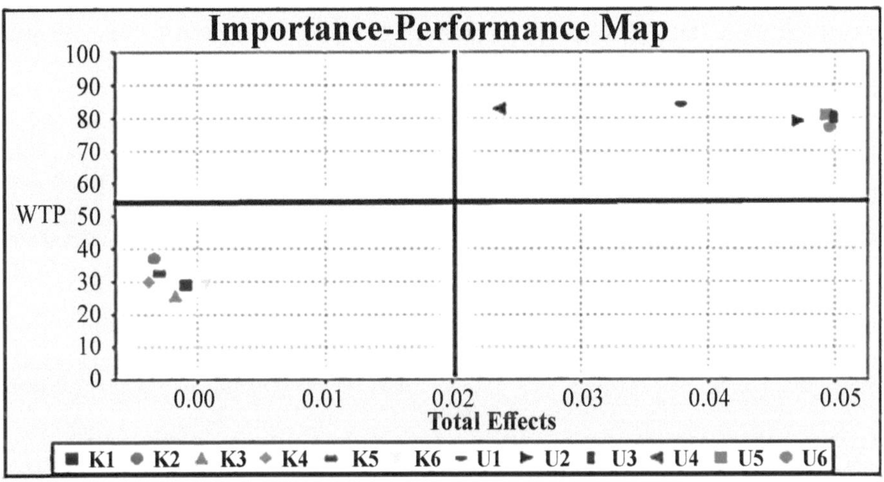

FIGURE 16.4 Item wise IPMA for knowledge and use.

Thus, if regulatory bodies and manufacturers try to increase the usage of food labels by making them more attractive, legible, and pictorial, it may lead to a positive increase in WTP, thereby improving the profit margin of the manufacturers as well.

KEYWORDS

- **food label information**
- **knowledge**
- **scale development**
- **SmartPLS**
- **structural model**
- **usage pattern**
- **willingness to pay**

REFERENCES

Affram, P. C., & Darkwa, S. (2015). Consumers' knowledge, understanding and use of food label information, and how it affects purchasing decision in Ho, Ghana.

Aziz, N. S., & Kamaludin, A., (2014). *Assessing Website Usability Attributes Using Partial Least*, *4*(2). https://doi.org/10.7763/IJIEE.2014.V4.423.

Brunsø, K., Grunert, K. G., & Fjord, T. A. (2002). *Consumers' food choice and quality perception* (Vol. 77). MAPP, Center for Marked Over Vågning,-Vurdering Og-Bearbejdning Til Fødevaresektoren.

Campos, S., Doxey, J., & Hammond, D., (2011). Nutrition labels on pre-packaged foods: A systematic review. *Public Health Nutrition*, *14*(8), 1496–1506. https://doi.org/10.1017/S1368980010003290.

Chiappa, G. D., (2015). *The Food Quality Labels: Awareness and Willingness to Pay in the Context of the Czech Republic*, *63*(2), 647–658.

Cowburn, G., & Stockley, L., (2014). *Consumer Understanding and Use of Nutrition Labeling: A Systematic Review.* https://doi.org/10.1079/PHN2004666.

Darkwa, S., (2016). *Consumers 'Knowledge, Understanding and Use of Food Label Information, and How it Affects Purchasing Decision in Ho, Information, and How it Affects Purchasing Decision in HO.*

Grunert, K. G., (2005). *Food Quality and Safety: Consumer Perception and Demand*, *32*(3), 369–391. https://doi.org/10.1093/eurrag/jbi011.

Hagmann, D., & Siegrist, M. (2020). Nutri-Score, multiple traffic light and incomplete nutrition labelling on food packages: Effects on consumers' accuracy in identifying healthier snack options. *Food Quality and Preference*, *83*, 103894.

Hair, J. F., Risher, J. J., Sarstedt, M., & Ringle, C. M., (2019). When to use and how to report the results of PLS-SEM. *European Business Review*, *31*(1), 2–24. https://doi.org/10.1108/EBR-11-2018-0203.

Henseler, J., Ringle, C. M., & Sarstedt, M., (2015). A new criterion for assessing discriminant validity in variance-based structural equation modeling. *Journal of the Academy of Marketing Science*, *43*(1), 115–135. https://doi.org/10.1007/s11747-014-0403-8.

Joel, D., Wayne, G., & Kathiravan, G., (2016). *Consumers 'Knowledge and Perception of Food Labels: The Case of Trinidad and Tobago*, *8*(11), 1130–1133.

Krystallis, A., & Ness, M., (2005). Consumer preferences for quality foods from a south European perspective: A conjoint analysis implementation on Greek olive oil. *International Food and Agribusiness Management Review*, *8*(2), 62–91.

Latvala, T., (2004). *Risk, Information and Trust in the Food Chain: Factors Explaining Consumer Willingness to Pay*, 691–705.

Miller, L. M. S., & Cassady, D. L., (2015). The effects of nutrition knowledge on food label use. A review of the literature. *Appetite*, *92*, 207–216. https://doi.org/10.1016/j.appet.2015.05.029.

Miškolci, S., (2011). *Consumer Preferences and Willingness to Pay for the Health Aspects of Food*, *LIX*(4), 167–176.

Nasreddine, L., Akl, C., Al-Shaar, L., Almedawar, M. M., & Isma'eel, H., (2014). Consumer knowledge, attitudes, and salt-related behavior in the middle-east: The case of Lebanon. *Nutrients*, *6*(11), 5079–5102. https://doi.org/10.3390/nu6115079.

Saxena, M., Sharma, M. K., & Jain, A., (2021). *Impact of Food Selection and Usage Pattern on Consumers' Attitude Towards Food Label Information*, *9*(1), 46–52. www.jpht.in (accessed on 6 December 2023).

Shmueli, G., Sarstedt, M., Hair, J. F., Cheah, J. H., Ting, H., Vaithilingam, S., & Ringle, C. M., (2019). Predictive model assessment in PLS-SEM: Guidelines for using PLSpredict. *European Journal of Marketing*, *53*(11), 2322–2347. https://doi.org/10.1108/EJM-02-2019-0189.

Van, R. W., & Frewer, L. J., (2008). Consumer perceptions of food quality and safety and their relation to traceability. *British Food Journal*, *110*(10), 1034–1046. https://doi.org/10.1108/00070700810906642.

Vemula, S. R., Gavaravarapu, S. M., Vardhana, V., Mendu, R., Mathur, P., & Avula, L., (2013). *Use of Food Label Information by Urban Consumers in India – a Study Among Supermarket Shoppers*, *17*(9), 2104–2114. https://doi.org/10.1017/S1368980013002231.

Vijaykumar, S., Lwin, M. O., Chao, J., & Au, C., (2013). Determinants of food label use among supermarket shoppers: A Singaporean perspective. *Journal of Nutrition Education and Behavior*, *45*(3), 204–212. https://doi.org/10.1016/j.jneb.2012.09.001.

Voon, J. P., Sing, K., & Agrawal, A., (2011). *Determinants of Willingness to Purchase Organic Food: An Exploratory Study Using Structural Equation Modeling*, *14*(2).

CHAPTER 17

Occupancy Monitoring to Prevent Spread of COVID-19 in Public Places Using AI

HARSHRIM PARDAL,[1] MANOHAR KAPSE,[2] VINOD SHARMA,[2] and N. ELANGOVAN[3]

[1]*Senior Cloud Analyst-HCM, Oracle Corporation, Bangalore, Karnataka, India*

[2]*Symbiosis Center for Management and Human Resource Development, Symbiosis International University, Pune, Maharashtra, India*

[3]*School of Business and Management, Christ University, Bangalore, Karnataka, India*

ABSTRACT

The chapter aims to automate the counting of people for occupancy monitoring and send an alert email if the occupancy exceeds the defined threshold in case of restricted occupancy guidelines. The study aims to reduce the manual error, effort, and time for people counting and provide a tool for footfall analysis. We propose and implement an occupancy monitoring system by counting the number of people entering and exiting a building/room using cameras and machine learning (ML) algorithms. The Single Shot Detector (SSD) algorithm, which is based on the MobileNet architecture, is used. This project provides an effective process for execution using either a recorded video file or a live stream from a camera. As the system automates counting people, it reduces human effort and error. It provides accurate results on time. The project can be implemented anywhere using a laptop and a camera for capturing the video. Thus, it provides high portability of the project. The system can leverage pre-installed CCTV cameras and systems in colleges,

Sustainability in Marketing Practice: Strategies for Industry 4.0.
Subrata Chattopadhyay, Udit Chawla, (Eds.)

malls, offices, etc. Thus, it requires less additional expenses and is economically friendly for the organization/decision-making authority. This chapter includes implications for various use cases such as ensuring adherence to COVID-19 guidelines by organizations, streamlining janitorial services, prevention of stampedes, improving indoor air quality, improving electricity efficiency, etc. This project fulfills an identified need to automate the people counting process and generate alerts accordingly.

17.1 INTRODUCTION

Severe acute respiratory syndrome (SARS–CoV-2) is a transmissible disease caused by novel Coronavirus (*Coronavirus*, n.d.). The first known case of COVID-19 was detected in December 2019 in Wuhan, China. The virus has high transmissibility because the virus from the mouth and nose of an infected person is thrown out through aerosol when a person cough or sneeze. The virus can also spread when the infected person speaks or breathe. The rate of spread of infection has made it easily reach every part of the world and qualified it as a pandemic.

Physical or social distancing, quarantining, covering of mouth and nose when coughing or sneezing, regular hand washing, ventilation of indoor spaces are examples of preventive methods for COVID-19. In public places, face masks or coverings have been advocated to reduce the risk of transmission. Since the outbreak began, The WHO has been collaborating with international experts, governments, and partners to expand scientific knowledge about Coronavirus rapidly, track the virus's spread and advise countries and individuals on how to protect their health and prevent the outbreak from spreading (COVID-19 Pandemic in India, 2020).

The first instances of COVID-19 infections were detected at three locations in Kerala, India on January 30th, 2020. State of emergency was declared in Kerala on March 23rd, 2020. Lockdown was announced for the entire country from March 25th, 2020. The number of daily infections peaked in the middle of September 2020, recording over 90,000 being affected every day. The rate of infections came down to a level of 15,000 during January 2021. Second wave of the pandemic began in March 2021. During the second wave, India reported over 40,000 new cases in a day and surpassed United States in terms of new and active cases (COVID-19 Pandemic in India, 2020).

Throughout the year 2020, several mutations of SARS-CoV-2 emerged. The first variant was named as Alpha, which is officially known as B.1.1.7 or UK variant (Davies, 2020). The second variant was named as Beta, referred to as B.1.351 (Tegally, 2021). The third variant known as Gamma is also referred to as P.1 (Faria, 2020). The fourth variant known as Delta and referred to as B.1.617.2 was more deadly and caused a major impact during the second wave (SARS-CoV-2 Variants of Concern and Variants under Investigation in England, Technical Briefing 19 Public Health England Technical Briefing 19, 2021).

As part of its fight against Coronavirus, nations have started vaccinating their citizens. On December 2nd, 2020, the first COVID-19 vaccination received regulatory approval from the UK medicines regulator MHRA. Emergency use authorization (EUA) status was provided for COVID vaccines by the US Food and Drug Administration (FDA) and several other nations. The new vaccines have been extensively credited with lowering the severity and fatality rates associated with COVID-19. On January 16th, India began administration of COVID-19 vaccines.

To curb and slow down the spread of the virus, Centers for Disease Control and Prevention (CDC) provides recommendations for taking various precautions in public spaces and multi-housing buildings (COVID-19 and Your Health, 2019). Though frequent disinfection and cleaning is recommended in places where there are movement of people, closeness, and longer period of exposure increases the chances of contacting virus. Personal protection strategies like social distancing, frequent handwash, wearing masks, cough, and sneeze etiquettes and quarantine were not effective when there is no crowd control in the public places or common utilities. Restricting the number of people in an area is required to create a safe congregate place. Scheduling and staggering the activity hours helps in serving a small group at a given time can reduce the close contact of people. For example, educational institutions across the globe proposed timetable consisting of alternate day of offline classes or hybrid classes. However, minimizing the traffic in common places such as walk ways, elevators, staircases or any covered places is needed to control the spread of virus.

The CDC also advises implementing social distancing during the COVID-19 pandemic by keeping a space of at least 6 feet between persons outside your home, not assembling in groups, and avoiding crowded places and public gatherings. These suggestions have an impact on whether an organization can safely restart operations. An office building, for example, could need to lower the number of people inside by roughly 11% to operate while practicing social distance, but a bar might need to cut its population by as much as 95% (Understanding the Impact of Social Distancing on Occupancy, 2021).

During the pandemic, nations across the globe have also imposed restrictions on occupancy in hotels, gyms, theaters, office workspaces, etc. The most common of such occupancy restrictions is placing a 50% occupancy restriction of people. Places such as hospitals, pharmacies, retail stores, malls, exhibition halls, religious places and public services are compelled to comply with governments regulations guidance and restrictions on the number of visitors at sites. Due to such restrictions, there is a need for an AI based systems to keep track of number of people at a place in real-time.

Occupancy monitoring means counting the number of people currently on the premises, or in a defined area. There are several ways to monitor occupancy, one of the most accurate is using CCTV or IP-enabled cameras and people counting devices. The live occupancy counts are shown on screens, on web browsers, smartphone apps and can be integrated into Internet-of-Things networks.

Overhead cameras are positioned over entries and exits. These connect to people counters which transmit the data over Wi-Fi, Internet, IoT or Ethernet to the display meters. Data can also be shown in dashboards on secure web browsers. Wide entrances and areas are handled by linking several cameras across the ceiling. A real time view of how many people is in each room, on each floor or in the entire building can be achieved. As well as live data the occupancy information is saved for trend analysis.

Occupancy monitoring systems provides solution to business to comply with government restrictions and maintain social distancing guidelines. The system supports automated decision-making for best utilization of space in terms of both effective workspace management in offices and customer management in public places. Accurate and real-time data of occupants is produced using software that is fed with data from smart sensors. These smart sensors include various types of devices like passive infra-red (PIR) sensor, Infrared array sensors, microwave sensors, people recognition cameras and ultrasonic sensors. Counting of the crowd in real-time is practiced in large performances, public shows, gig work settings, forbidden areas in manufacturing units, traffic controls, etc., to ensure safety and minimize health risks (Sassisegarane, 2021). Counting crowds is also done in automating resource allotment, counting attendance, surveillance, retail store footfall analysis (Khan et al., 2020; Retail Footfall Analytics Solution | Store Customer Data Insights | LTI, 2021).

This chapter describes the development of an occupancy monitoring system which will process the video in real time and tracks people entering or exiting the stores/buildings/shopping malls, etc., and provides the number of people currently in the area. As governments around the globe have imposed occupancy restrictions in view of the COVID-19, the system is a step in the direction of adhering to the guidelines by sending an alert to the staff if the people are way over the threshold limit. This system can also act as a measure towards footfall analysis and crowd analysis.

17.2 REVIEW OF LITERATURE

There are lot of studies which had been done in the field of occupancy monitoring using AI, ML, and vision analytics. Mangawati et al. (2018) identified the best algorithms for various types of tracking through an extensive survey. They compared various object tracking algorithms used in different environmental conditions to identify the efficient algorithms. Gowari et al. (2021) prescribed a system for tracking humans and detecting face masks using convolution neural network (CNN) and MobileNet SSD. CNN is an image classification algorithm. MobileNet is a neural network that is used for classification and recognition (Howard et al., 2017). Whereas the SSD is a framework that is used to realize the multi-detector (Gowari et al., 2021).

Kleminger et al. (2013) used the electricity consumption data to detect the occupant's presence. A classification model was trained using support vector machine (SVM) on the data gathered from smart electricity meters of five households. K-nearest neighbor (KNN), thresholding (THR) and hidden Markov model (HMM) were also used.

Saha et al. (2019) mentioned about three different types of algorithms for occupancy detection, namely, detection algorithms, counting algorithms and tracking algorithms. Presence of an occupant in an area of a building is found using detection algorithms. For this KNN or SVM are used. Saha et al. (2019) compared the different methods according to their time and cost for deployment, robustness, and model flexibility. They found that probabilistic graphical modeling (PGM) kernel-based methods, neural networks (NNs), data mining, clustering, statistical, and classical methods had least time and cost to deploy. They also provided high robustness and model flexibility.

Shih (2014) proposed an occupancy monitoring system which combines 'vision based and sensor-based cameras.' It also discussed Gaussian mixture models for occupancy monitoring and outlined the challenge of overcoming human object detection confusion when the humans are overlapped or occluded. Balci et al. (2018) stated that single shot detection (SSD) algorithm is one of the effective object detection methods by speed and accuracy. They detect the presence of human using deep learning-based object detection techniques, spatial elimination and classification stages.

SSD is a multi-category single-shot detector that is faster and accurate than the previous ones like YOLO, which uses explicit region suggestions and pooling. Liu et al. (2016) also spoke about the accuracy of SSD in their work and have concluded that SSDs have high accuracy at the object proposal phase and is considerably faster. They also provide a unified framework for both training and inference. SSDs are much more accurate for smaller input image size when compared to single stage methods.

Bashir et al. (2020) proposed an inexpensive standard operating procedure (SOP) compliance system for COVID protocols that can counts the number of persons entering and departing an area. Using the Internet of Things (IoT), they also provided alerts to attendees and managers on the physical distance and body temperature. The proposed system used infrared obstacle detectors placed at the entry and exit points. It also provided the number of people per square meter. A wide-angle camera module on a raspberry Pi (RPi) was utilized to identify any violations in the minimum distance between queued individuals. Using OpenCV, MobileNet-SSD, and the Caffe model, Identification of individuals, classification, and bounding box localization were performed. However, this solution was only useful in checking violation of social distance in a specifically monitored queue.

Feng et al. (2019) constructed a comprehensive human head detection system. Image feature extraction was done using the lightweight network MobileNet. The SSD was adopted in the lower layer with the activation function SeLU. Their experiments proved that the use of network based MobileNet produced a higher detection accuracy. It also had a lower latency, better robustness, and was effective on number of model parameters.

Fernández-Caramés et al. (2020) estimated the occupancy level using a novel IoT. This occupancy system allows estimation of people in public spaces including public transport systems like buses and trains in real time. The number of people in a particular place was tracked using the wearable devices worn by each occupant. The work included estimating the number of people in a room by tracking the wearable worn by each occupant. Sathyabama et al. (2020) monitored the real-time social distancing and face mask detection using a deep learning framework. This study adopts MobileNetV2 for masked/no masked datasets. An accuracy of approximately 98% was achieved. Considering the previous studies SSD is the best for the given scenario for occupancy monitoring.

17.2.1 OBJECTIVES

- To count the number of occupants in the buildings, stores, colleges, shopping malls, etc., on a real-time basis.
- Enable alerting the organization by sending an alert mail to the staff if the people are way over the limit.
- Act as a measure towards footfall analysis and in a way to tackle COVID-19.

17.2.2 TOOLS USED

Python is a high-level, interpreted programing language (Van Rossum & Drake, 2009). In this project, Python is adopted as the programming language. The libraries used in this project are NumPy (Harris et al., 2020), ssl (S.S.L., n.d.), csv (Van Rossum & Drake, 2009), datetime (Datetime, 2022), cv2 (Bradski, 2000), time (Time – Time Access and Conversions — Python 3.9.1 Documentation, n.d.), schedule (Schedule, 2022), argparse (Argparse, 2022), imutils (Imutils, 2022), dlib (Dlib, 2022), scipy.spatial (Virtanen et al., 2020), smtplib (Smtplib — SMTP Protocol Client — Python 3.9.2 Documentation, 2022), OpenCV Python (Bradski & Kaehler, 2008), Microsoft Excel (Microsoft, 2021).

17.2.3 TECHNIQUES ADOPTED

An SSD (Single Shot Detector) with a MobileNet architecture is being used. In general, SSD only takes a single shot to detect whatever is in an image, SSD seasoned with a MobileNet should result in a faster and more efficient object detector (Liu et al., 2016). MobileNet, is a DNN designed to run on resource constrained devices (Howard et al., 2017). Centroid tracker is one of the most reliable trackers (Simple Object Tracking with OpenCV, 2021).

17.3 METHODOLOGY (FIGURE 17.1)

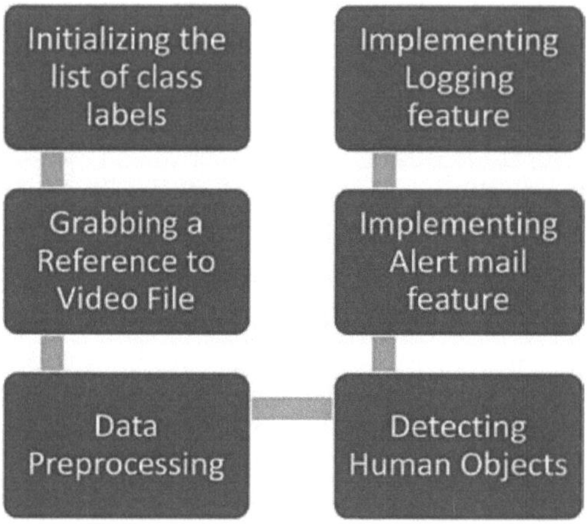

FIGURE 17.1 Process adopted.

17.3.1 INITIALIZING THE LIST OF CLASS LABELS

To initialize the model class labels such as human, boat, bottle, airplane, car, etc., were initialized using a python list.

17.3.2 GRABBING A REFERENCE TO THE VIDEO FILE

Next, a reference to the input video file is grabbed. The input video can be either through a webcam or through a specific video file stored in the system. The input is taken using the VideoCapture method of the cv2 package in python. Frame Dimensions are specified, and FPS estimator is started using the FPS library of the imutils package.

17.3.3 DATA PREPROCESSING

Data preprocessing in this chapter involves feature engineering of the input video. The capability of feature engineering is leveraged through the in-built provided capabilities of the single shot detector. SSD uses VGG-16 as its base network. In SSD fully connected layers, a set of *auxiliary* convolutional layers are added. These layers enable feature extraction at multiple scales and progressively decreases the size of the input to each subsequent layer (Forson, 2019) (Figures 17.2 and 17.3).

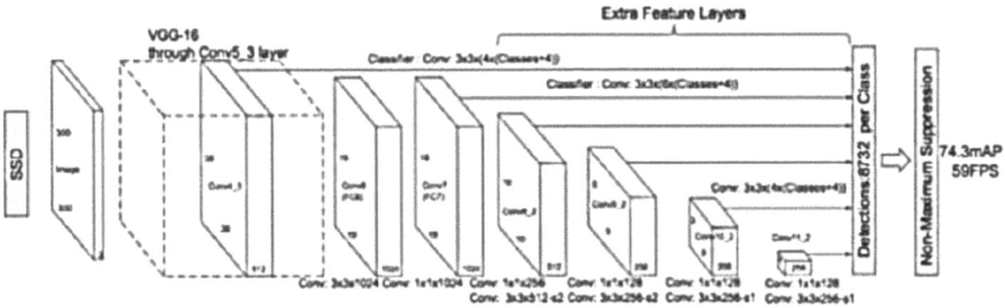

FIGURE 17.2 Architecture of SSD.

Source: Reprinted from Liu et al., (2016).

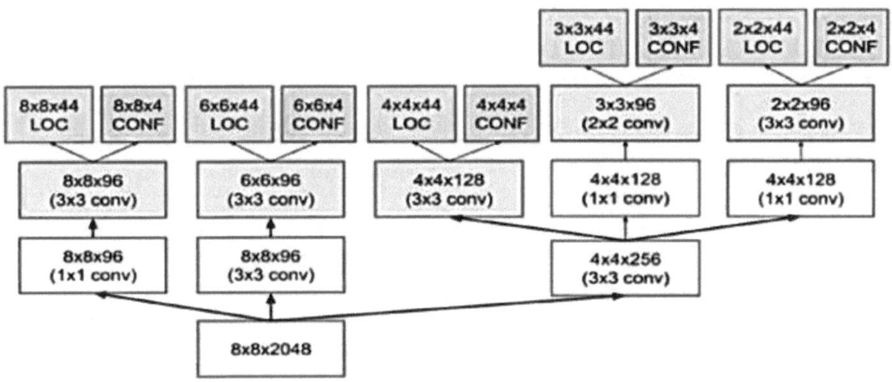

FIGURE 17.3 Architecture of multi-scale convolutional prediction of the location and confidences of multibox.

SSD discretizes the output space of bounding boxes into a set of default boxes over different aspect ratios and scales per feature map location. Small convolutional filters to predict object categories and offsets in bounding box locations, separate predictors (filters) are used for different aspect-based detections. The 1×1 convolutions in the above architecture aid in dimensionality reduction. The resulting architecture has 11 priors per feature map cell (8×8, 6×6, 4×4, 3×3, 2×2) and only one on the 1×1 feature map, totaling 1,420 priors per image, allowing for robust coverage of input images at multiple scales and the detection of objects of various sizes (Forson, 2019). MultiBox only keeps the top K predictions that have the fewest location (LOC) and confidence (CONF) losses finally. This is followed by hard negative mining which helps in reducing the imbalance between positive and negative training examples. The negative examples are sorted using the highest CONF loss for each default box and pick the top ones so that the ratio between the negatives and positives is at most 3:1. This is followed by data augmentation which helps make the model more robust to various input object sizes and shapes, each training image is randomly sampled. This is achieved through the dnn library of the cv2 package.

This is followed by detecting human objects which is done by converting the frame to a blob and passing the blob through the network. The blobFromImage method of the cv2 package is used for the same.

17.3.4 IMPLEMENTING THE ALERT MAIL FEATURE

An email alert in real-time is created using smptlib and ssl package. An alert mail is triggered and sent to the registered email id if the total number of people exceeds the threshold level specified for the store/building.

17.3.5 IMPLEMENTING THE LOGGING FEATURE

This feature enables collection of logs at end of the video/day. This feature is implemented using the datetime and csv packages. It stores data in the form of a csv file which gets stored automatically on the system and is useful for footfall analysis (Figure 17.4).

FIGURE 17.4 The project execution in tracking state.

17.4 DISCUSSION AND CONCLUSION

With the help of SSD and MobileNet architecture, a model is built to count the number of people in a building and accordingly send alert emails if the number of people currently in the building exceeds the threshold specified in an efficient way. SSD helps in overcoming the bottlenecks of R-CNN such as slow network at inference time (i.e., when dealing with training data), training the data is a lengthy process and unwieldy. This, in turn, reduces the manual effort, provides an opportunity for timely responses to situations of concern, and reduces human error and human effort. The system can leverage the already pre-installed CCTV cameras and systems in colleges, malls, offices, etc. Thus, it requires fewer additional expenses and is economically friendly for the organization/decision-making authority. The system also provides an output in the form of a log file which can prove to be a crucial step in footfall analysis and also helps in sending real-time alert emails in case of occupancy exceeding the set threshold values. This system helps in stopping the spread of COVID in real-time. It helps in keeping a regular disinfection and cleaning schedule (Five Key Benefits of People Counting & Occupancy Detection, 2021). It is critical to maintain public order in congested areas such as airports, carnivals, sporting events, and train stations. Counting people is an important part of a crowd control system and can prevent stampedes also (Tech That: Size up a Crowd the Neural Way, 2017).

17.4.1 LIMITATIONS

The model runs well on well-lit videos. The input video for the model should be a top-down view of the entrance point of a room/building. Due to computational restrictions, the project was executed by running the model on an input video of approximately 2 minutes.

17.4.2 FUTURE WORK

The project can be enhanced for running on separate cameras for entry and exit points. The project can be enhanced by connecting the output to a digital display which will show the real-time count of people in the building. The project can be enhanced for sending the alert email on more than one email ID.

KEYWORDS

- **alert**
- **alert mail feature**
- **data preprocessing**
- **logging**
- **MobileNet**
- **occupancy monitoring**
- **single shot detector**

REFERENCES

Argparse, (2022). *Parser for Command-Line Options, Arguments and Sub-commands* — Python 3.10.1 Documentation. https://docs.python.org/3/library/argparse.html (accessed on 6 December 2023).

Balci, B., Alkan, B., Elihos, A., & Artan, Y., (2018). Front seat child occupancy detection using road surveillance camera images. In: *2018 25th IEEE International Conference on Image Processing (ICIP)*. https://doi.org/10.1109/icip.2018.8451582.

Bashir, A., Izhar, U., & Jones, C., (2020). IoT based COVID-19 SOP compliance monitoring and assisting system for businesses and public offices. *Proceedings of 7th International Electronic Conference on Sensors and Applications*. https://doi.org/10.3390/ecsa-7-08267.

Bradski, G., & Kaehler, A., (2008). *Learning OpenCV: Computer Vision with the OpenCV Library.* O' Reilly Media, Inc.

Bradski, G., (2000). The OpenCV library. *Dr. Dobb; Journal of Software Tools*.

Coronavirus. (n.d.). https://www.who.int/health-topics/coronavirus#tab=tab_1 (accessed on 6 December 2023).

COVID-19 and Your Health. (2019). In *Centers for Disease Control and Prevention*. https://www.cdc.gov/coronavirus/2019-ncov/hcp/infection-control-recommendations.html.

COVID-19 Pandemic in India. (2020). https://en.wikipedia.org/wiki/COVID-19_pandemic_in_India (accessed on 6 December 2023).

Datetime. (2022). *Basic Date and Time Types* — Python 3.9.5 Documentation. https://docs.python.org/3/library/datetime.html (accessed on 6 December 2023).

Davies, N., (2020). *Estimated Transmissibility and Impact of SARS-CoV-2 Lineage B.1.1.7 in England*. https://cmmid.github.io/topics/covid19/uk-novel-variant.html (accessed on 6 December 2023).

Dlib. (2022). *dlib*. https://pypi.org/project/dlib/18.17.100/ (accessed on 6 December 2023).

Faria, N. R., (2020). Genomic characterization of an emergent SARS-CoV-2 lineage in Manaus: Preliminary findings. *Virological*. https://virological.org/t/genomic-characterisation-of-an-emergent-sars-cov-2-lineage-in-manaus-preliminary-findings/586 (accessed on 6 December 2023).

Feng, X., Xie, R., Sheng, J., & Zhang, S., (2019). Population statistics algorithm based on MobileNet. *Journal of Physics: Conference Series*, *1237*(2), 22045. https://doi.org/10.1088/1742-6596/1237/2/022045.

Fernández-Caramés, T., Fraga-Lamas, P., & Froiz-Míguez, I., (2020). An IoT and blockchain based system for monitoring and tracking real-time occupancy for COVID-19 public safety. *Proceedings of 7th International Electronic Conference on Sensors and Applications*. https://doi.org/10.3390/ecsa-7-08207.

Five Key Benefits of People Counting & Occupancy Detection, (2021). https://behrtech.com/blog/5-key-benefits-of-people-counting-occupancy-detection/ (accessed on 6 December 2023).

Gowari, Y., Gaikwad, R., Gurav, A., & Raut, V., (2021). People monitoring and mask detection using real-time video analyzing. *VIVA-Tech International Journal for Research and Innovation*, *1*(4).

Harris, C. R., Millman, K. J., & Walt, S. J., (2020). Array programming with NumPy. *Nature*, *585*, 357–362. https://doi.org/10.1038/s41586-020-2649-2.

Howard, A. G., Zhu, M., Chen, B., Kalenichenko, D., Wang, W., Weyand, T., Andreetto, M., & Adam, H., (2017). *MobileNets: Efficient Convolutional Neural Networks for Mobile Vision Applications*. CoRR, abs/1704.0. http://arxiv.org/abs/1704.04861.

Imutils. (2022). Imutils. https://pypi.org/project/imutils/ (accessed on 6 December 2023).

Khan, K., Albattah, W., Khan, R. U., Qamar, A. M., & Nayab, D., (2020). Advances and trends in real time visual crowd analysis. *Sensors*, *20*(18), 5073. https://doi.org/10.3390/s20185073.

Kleminger, W., Beckel, C., Staake, T., & Santini, S., (2013). Occupancy detection from electricity consumption data. *Proceedings of the 5th ACM Workshop on Embedded Systems for Energy-Efficient Buildings, Build Sys.*, *1*, 1–8.

Liu, W., Anguelov, D., Erhan, D., Szegedy, C., Reed, S., Fu, C.-Y., & Berg, A. C. (2016). *SSD: Single Shot MultiBox Detector* (pp. 21–37). https://doi.org/10.1007/978-3-319-46448-0_2.

Mangawati, A., Mohana, L. M., & Aradhya, H. V., (2018). Object tracking algorithms for video surveillance applications. In: *2018 International Conference on Communication and Signal Processing (ICCSP)*. https://doi.org/10.1109/iccsp.2018.8524260.

Microsoft, (2021). *Microsoft Excel-Spreadsheet Software* | Microsoft 365. https://www.microsoft.com/nl-nl/microsoft-365/excel?legRedir=true&CorrelationId=9d28b09f-3480-4d19-b106-ac3375820bf2&rtc=1 (accessed on 6 December 2023).

Retail Footfall Analytics Solution | Store Customer Data Insights | *LTI*, (2021). https://www.lntinfotech.com/industries/cpg-retail/cpg/retail-footfall-analytics/ (accessed on 6 December 2023).

S.S.L. (n.d.). *TLS/SSL Wrapper for Socket Objects* — Python 2.7.18 Documentation. https://docs.python.org/2/library/ssl.html (accessed on 6 December 2023).

Saha, H., Florita, A. R., Henze, G. P., & Sarkar, S., (2019). Occupancy sensing in buildings: A review of data analytics approaches. *Energy and Buildings*, *188*, *189*, 278–285. https://doi.org/10.1016/j.enbuild.2019.02.030.

S*ARS-CoV-2 Variants of Concern and Variants Under Investigation in England, Technical Briefing 19 Public Health England Technical Briefing*, *19*, (2021). https://assets.publishing.service.gov.uk/government/uploads/system/uploads/attachment_data/file/1005517/Technical_Briefing_19.pdf (accessed on 6 December 2023).

Sassisegarane, P., (2021). *Dense and Sparse Crowd Counting Methods and Techniques: A Review*. https://medium.com/nanonets/dense-and-sparse-crowd-counting-methods-and-techniques-a-review-bae04fdbf062 (accessed on 6 December 2023).

Sathyabama, B., Devpura, A., Maroti, M., & Rajput, R. S., (2020). Monitoring pandemic precautionary protocols using real-time surveillance and artificial intelligence. In: *2020 3rd International Conference on Intelligent Sustainable Systems (ICISS)*. https://doi.org/10.1109/iciss49785.2020.9315934.

Schedule. (2022). Schedule. https://pypi.org/project/schedule/ (accessed on 6 December 2023).

Shih, H., (2014). A robust occupancy detection and tracking algorithm for the automatic monitoring and commissioning of a building. *Energy and Buildings*, *77*, 270–280. https://doi.org/10.1016/j.enbuild.2014.03.069.

Simple Object Tracking with OpenCV, (2021). https://www.pyimagesearch.com/2018/07/23/simple-object-tracking-with-opencv/ (accessed on 6 December 2023).

Smtplib — SMTP Protocol Client — Python 3.9.2 Documentation, (2022). https://docs.python.org/3/library/smtplib.html (accessed on 6 December 2023).

Tech That: Size up a Crowd the Neural Way, (2017). https://www.deccanchronicle.com/decaf/030917/tech-that-size-up-a-crowd-the-neural-way.html (accessed on 6 December 2023).

Tegally, H., (2021). *Nature*. 10.1038/s41586-021-03402-9.

Time – Time access and conversions – Python 3.9.1 documentation. (2022). https://docs.python.org/3/library/time.html (accessed on 6 December 2023).

Understanding the Impact of Social Distancing on Occupancy, (2021). https://www.usfa.fema.gov/coronavirus/planning_response/occupancy_social_distancing.html (accessed on 6 December 2023).

Van, R. G., & Drake, F. L., (2009). *Python 3 Reference Manual*. CreateSpace.

Virtanen, P., Gommers, R., Oliphant, T. E., Haberland, M., Reddy, T., Cournapeau, D., Burovski, E., Peterson, P., Weckesser, W., Bright, J., Walt, S. J., Brett, M., Wilson, J., Jarrod Millman, K., Mayorov, N., Nelson, A. R. J., Jones, E., Kern, R., Larson, E., & Contributors. (2020). SciPy 1.0: Fundamental Algorithms for Scientific Computing in Python. *Nature Methods*, *17*(3), 1 0. https://doi.org/10.1038/s41592-019-0686-2.

Index

D

N

O

T